Eiichi Baba, Hideo Kawarada, Wataru Nishijima,
Mitsumasa Okada, Hiroshi Suito

Waves and Tidal Flat Ecosystems

Springer-Verlag Berlin Heidelberg GmbH

E. Baba, H. Kawarada, W. Nishijima,
M. Okada, H. Suito

Waves and Tidal Flat Ecosystems

With 144 Figures, 27 in colour

Springer

PROF. EIICHI BABA
Information Exchange Office
Hiroshima University
1-3-2 Kagamiyama,
Higashi-Hiroshima, 739-8511
Japan

PROF. HIDEO KAWARADA
Faculty of Distribution and
Logistics
Systems
Ryutsu Keizai University
Hirahata 120, Ryuugasaki
Ibaraki, 301-8555
Japan

PROF. WATARU NISHIJIMA
PROF. MITSUMASA OKADA
Department of Chemical Engineering
Graduate School of Engineering
Hiroshima University
1-4-1 Kagamiyama
Higashi-Hiroshima, 739-8527
Japan

PROF. HIROSHI SUITO
Department of Environmental
and Mathematical Sciences
Okayama University
3-1-1 Tsushima-naka
Okayama, 700-8530
Japan

ISBN 978-3-642-62444-5 ISBN 978-3-642-55534-3 (eBook)
DOI 10.1007/978-3-642-55534-3

Cataloging-in-Publication Data applied for

A catalog record for this book is available from the Library of Congress.

Bibliographic information published by Die Deutsche Bibliothek
Die Deutsche Bibliothek lists this publication in the Deutsche Nationalbibliografie; detailed bibliographic data is available in the Internet at <http://dnb.ddb.de>.

http://www.springer.de

Originally published by Springer-Verlag Berlin Heidelberg New York in 2003
Softcover reprint of the hardcover 1st edition 2003

Camera ready by the authors
Cover design: E. Kirchner, Heidelberg
30/3140/ 5 4 3 2 1 0

Preface

In this book, the authors address present-day methodology to explore natural phenomena. Highly advanced computational fluid mechanics contributes to the rationalization of experimental work. By taking sandy beach and tidal flat ecosystems as an example, results from interdisciplinary collaboration between environmental experimentalists and applied mathematicians are presented in an interactive manner.

In 1994, an environmental study group from Hiroshima University started a study on tidal flats. Based on fieldwork, they discovered the importance of fluid flow in understanding the biological activities of tidal flats and sandy beaches. They found that the number of bacteria on the seabed strongly correlates with the amount of silt (fine sand < 50 micron). Larger slopes produce smaller amounts of silt. They noted the importance of an appropriate slope for a tidal flat where sound biological activities are carried out on the seabed.

Furthermore, independently in 1993, an applied mathematics group from Chiba University studied a sliding problem of two different media. In 1995, they extended their work to explain a penetration problem from one side to the other in different phases. Then, in 1996, the study was extended to wave motion on the sloping sandy beach. At this moment this mathematics group did not exploit yet their potential to explore coastal environmental problems.

In 1997, the environmental group started a research project dealing with the effect of spilled oil on coastal ecosystems. This project aims to provide measures to counter environmental disasters such as the spilled oil in 1997 from a tanker in the Japan Sea. In this study, a new experimental discovery was made about the role of waves over a sandy beach. Breaking waves act as a pump to transport fresh seawater to the beach. Thus, it became possible to estimate quantitatively the infiltration of seawater by waves into the seabed.

Even though the waves themselves are only a few centimeters high, the importance of waves was recognized from experimental studies using a tidal flat sumulator, which contains actual soil from the tidal flat. That is, benthos grows in the seabed when waves and tides act together over the tidal flat. However, the number of benthos decreases and sea grass grows instead when the tide alone acts without waves. Sea grass prevents the infiltration of fresh seawater into the seabed. Thus, it was clear that different kinds of organisms flourish under different habitat conditions, i.e., with waves or without waves. Thus waves at the margin of sandy beaches or tidal flats are an important factor for the determination of coastal ecosystems.

The environmental experiment group needed rationalization of their experimental results from a theoretical standpoint. The Chiba University mathematics group was informed of the flow phenomenon observed in experiments by one of the environmental study groups in 1998. These experimental results attracted and inspired the mathematics group to exploit the large potential to explore the flow phenomenon inside the seabed. They then extended their analysis to treat fluid flow in various environments such as air, wet sand and dry sand. It was found that mathematically simulated flow behaviour inside a sloping beach well explained observed phenomena. Thus, the flow phenomena observed by the einvironmented study group were rationalized.

The einvironmental study group continued their work and found that spilled oil over a sloping beach prevents the infiltration of fresh seawater into the seabed, causing a shortage of oxygen. Therefore, the benthos suffers.

The mathematics group extended their work to simulate bahaviour and decomposition of spilled oil based on Navier-Stokes equations with the Bingham fluid model for oil. Multi-phase flow analyses were made. In this study, decomposition of spilled oil by bacteria is simulated as a chemical reaction. Thus, natural phenomena around the margin of land, in other words, the margin of different disciplinary research fields are being explored by collaboration of experimentalists and applied mathematicians. The development of this work in the future is highly expected to further our understanding of coastal ecosystems and to contribute to sound engineering approaches in the construction of artificial tidal flats and coastlines.

March 2003
Eiichi Baba
Hideo Kawarada
Mitsumasa Okada

The following authors contributed to this book:

Chapter 1; W. Nishijima
Chapter 2; E. Baba
Chapter 3; H. Kawarada and H. Suito
Chapter 4; M. Okada
Chapter 5; H. Kawarada and H. Suito
Chapter 6; M. Okada
Chapter 7; H. Kawarada and H. Suito
Chapter 8; H. Kawarada and H. Suito
Chapter 9; H. Kawarada and H. Suito

Table of Contents

1 What is tidal flat?

1.1 Introduction

Wave and current energies are important factors in determining the physico-chemical and biological conditions in coastal ecosystem. Sandy beach develops in the area of very high physical energies facing open ocean where coarse grained sands make beach and sediments are completely oxidized. On the other hand, tidal flat develops under sluggish currents in river mouth and semi-enclosed coastal area where finer particles are trapped in the flat and sediments show strong reducing conditions sometimes to sediment surface.

Tidal flat ecosystem differs from the aquatic ecosystem in the sense that the former is exposed to sunlight directly at ebb tide and is submerged at flood tide. In tidal flat the primary producers are not phytoplanktons but phytobenthos. Pelagic production such as production by phytoplanktons is also supplied from seawater. Filter feeding benthic organisms such as bivalvia depend on currents to transport oxygen and pelagic productions and carry away wastes. There are other benthic organisms, which move through the sediment and take food from the sediment itself. They are called deposit feeders and include worms and amphipods. Tidal flat ecosystem is not a uniform system. Physico-chemical gradients are observed along the depth such as oxidizing and reducing conditions in the sediment, fluid condition and salinity in overlaying water. These variations produce rich benthic systems.

As a result of industrial and urban developments of coastal area in Japan, tidal flats are endangered. Like natural wetlands in U.S.A, the loss of tidal flats is one of the largest environmental concerns in Japan. According to the Environmental Agency in Japan, the area of natural tidal flats was approximately 82,600 ha during the 1940s. However, by the 1980s, nearly forty percent of the natural ones were lost (Kimura 1994). Currently, many efforts have been made to protect natural tidal flats and wetland ecosystems. In addition, there were many projects aiming to restore damaged tidal flats and create man-made tidal flats to mitigate lost ones (Confer et al. 1992, Miyoshi et al. 1990, Ogura et al. 1995).

Any creation and restoration of tidal flats should recover functions and values of natural tidal flats. Despite increasing restoration and creation projects, only a few studies have been carried out on the functions of man-made ones. For man-made tidal flats to secure the typical functions of natural tidal flats, it is necessary to identify and understand relationships between functions and environmental condition, design criteria, and age of man-made tidal flats.

This chapter deals with the experimental approach to reveal the function and structure of both natural and man-made tidal flats and the factors to determine the function and structure of tidal flat through on-site investigations and laboratory experiments. (Lee J.G et al. 1997, 1998a, 1998b, 1999a, 1999b).

1.2 Function and structure of natural and man-made tidal flats

1.2.1 Physico-chemical characteristics

The study sites were selected in the northern part of Hiroshima Bay, which is located in the western part of the Seto-Inland Sea, Japan (Fig.1.1). The Ohta River flows into the northern part of this bay. Table 1.1 summarized characteristic features of natural and man-made tidal flats. Man-made tidal flats were constructed for various purposes like oyster and clam cultures, bathing resort and compensatory mitigation. The significant difference in water quality in the northern part of Hiroshima Bay was not observed: DO (dissolved oxygen) and COD (chemical oxygen demand) at St.2 located at the middle of the bay and St.8 located at the back of the bay ranged from 10.7 to 12.4 mg l^{-1} and from 0.9 to 1.2 mg l^{-1}, respectively (Maritime Safety Agency 1997). It was presumed that the water quality in the northern part of the bay and, offshore of all the tidal flats was almost the same.

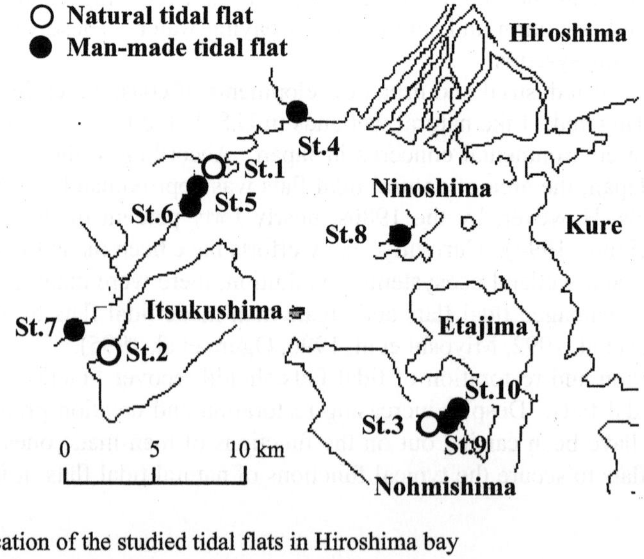

Fig. 1.1. Location of the studied tidal flats in Hiroshima bay

Table 1.1. The studied tidal flats in Hiroshima bay

Site	Origin	Area (ha)	Slope	Constructed year	Purpose
St.1	Natural	24	1.0/100	-	
St.2	Natural	4.7	1.0/100	-	
St.3	Natural	3.3	1.6/100	-	
St.4	Man-made	24	2.5/100	1990	Compensatory mitigation
St.5	Man-made	3.2	0.4/100	1986	Oyster culture
St.6	Man-made	6.7	0.7/100	1973	Oyster culture
St.7	Man-made	1.2	1.5/100	1984	Clam culture
St.8	Man-made	5.0	2.0/100	1987	Clam culture
St.9	Man-made	2.3	4.0/100	1987	Bathing resort
St.10	Man-made	2.3	0.7/100	1987	Oyster culture

Table 1.2. Physico-chemical characteristics of sediment in the tidal flats

Site	Origin	Mean medium (mm)	Silt (%)	Carbon (mg g-dry^{-1})	Nitrogen (mg g-dry^{-1})	Reducing zone (cm)
St.1	Natural	0.49	4.77	10.25	0.93	-2
St.2	Natural	0.39	2.71	3.30	0.33	-2
St.3	Natural	0.57	8.68	5.28	0.52	-2
St.4	Man-made	1.60	0.03	1.50	0.07	ND
St.5	Man-made	0.88	0.85	1.28	0.13	-25
St.6	Man-made	0.42	0.35	1.52	0.10	-15
St.7	Man-made	0.49	0.50	2.08	0.23	-20
St.8	Man-made	0.40	0.87	0.90	0.10	ND
St.9	Man-made	0.80	0.05	1.09	0.10	ND
St.10	Man-made	0.46	2.78	3.78	0.30	-2

Reducing zone: soil depth from the surface showing minus potential.
ND: reducing zone was not detected within core sample of 30 cm.

Fig.1.2 shows vertical profiles of particle size distribution of soils in the typical natural (St.2) and man-made (St.8) tidal flats. Median particle diameters in the natural tidal flat were 0.6 mm in the surface (0 ~ 2 cm) and 0.28 ~ 0.42 mm under 2cm-depth. Median particle diameters in the man-made tidal flat were 0.26 ~ 0.38 mm through all depths. There is no significant difference between the two tidal flats with respect to the median particle diameter. However, silt content indicated large difference between the tidal flats. The natural tidal flat had larger amount of silt (3.5 ~ 6.6 %) compared with the man-made tidal flat (0.4 ~ 1.6 %). The contents of silt for all tidal flats studied are given in Table 1.2. The silt contents in the natural tidal flats were higher than those in the man-made tidal flats except St.10. The silt content at St.10 was the highest among the man-made tidal flats and was almost the same as those in the natural ones. The organic carbon content of soil was in the range of 3.30 to 10.25 mgC g^{-1} in the natural tidal flats and 0.90 to 3.78 mgC g^{-1} in the man-made ones. The organic carbon at St.10 was 3.78 mgC g^{-1}, which was also the highest among the man-made ones and was also similar to those in the natural ones.

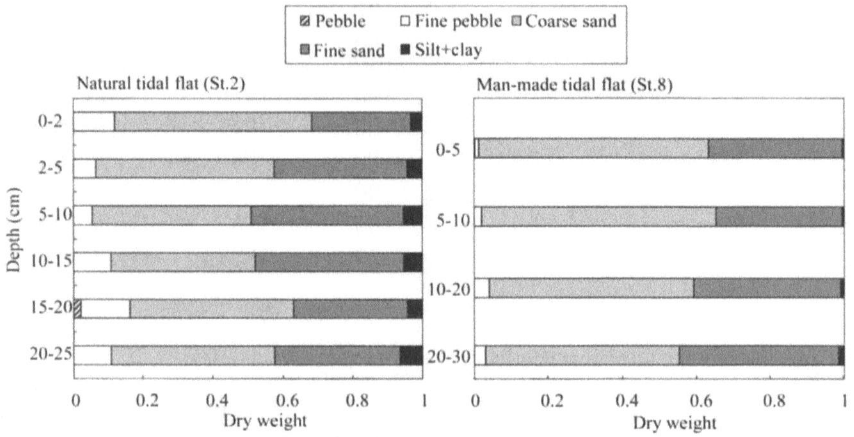

Fig. 1.2. Vertical profiles of particle size distribution of soils in natural and man-made tidal flats. Pebble: >4.75mm, Fine pebble: 2.00~4.75 mm, Coarse sand: 0.425~2.00 mm, Fine sand: 0.075~0.425 mm, Silt+sand: 0~0.075 mm

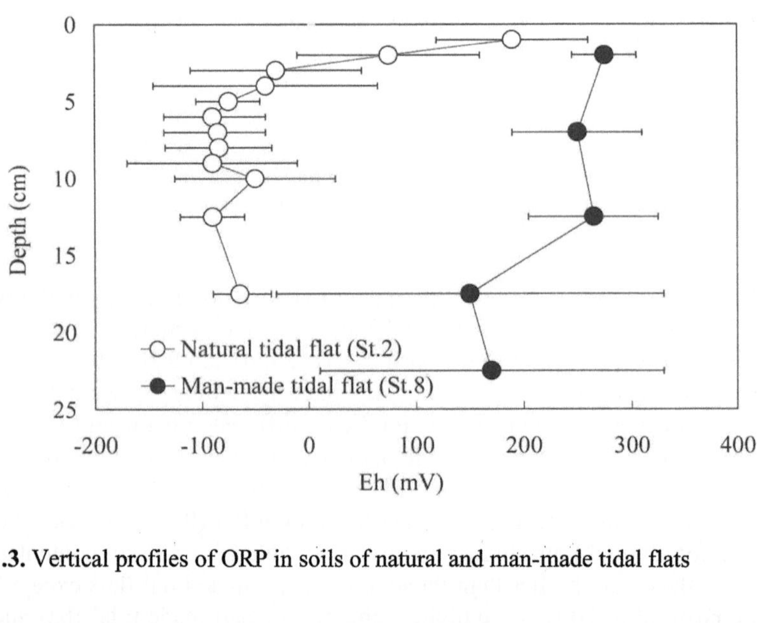

Fig. 1.3. Vertical profiles of ORP in soils of natural and man-made tidal flats

Vertical profiles of ORP (oxidation-reduction potential) at St.2 and St.8 are shown in Fig.1.3. The ORP of soil at St.2 showed positive values from the surface to 4cm in depth and dropped to negative values at zones below 4 cm but negative values were not noted through all depth studied. Other man-made tidal flats except St.10 also showed oxidizing zone in soil shallower than 15 cm in depth. It is clear that there were significant differences in the contents of silt, organic carbon and vertical profile of ORP between the natural and man-made tidal flats. However, the soil at St.10 showed almost the same physico-chemical characteristics with those in the natural ones even though St.10 is a man-made tidal flat.

1.2.2 Hydraulic characteristics

Fig.1.4 shows the diurnal changes of groundwater levels in the natural (St.2) and man-made (St.8) tidal flats. The groundwater level at St.2 did not fall beyond 2 cm even though the seawater level was 135 cm lower than the site studied. On the other hand, the groundwater level at St.8 decreased until 9 cm in depth. These results indicate that seawater and air can be induced into soil every tidal action until 2 cm and 9 cm in depth in the St.2 and St.8, respectively, corresponding to the results of ORP. Those may be the active zones of bacteria and benthic organisms and the responsible zones for the degradation of organic matter in seawater.

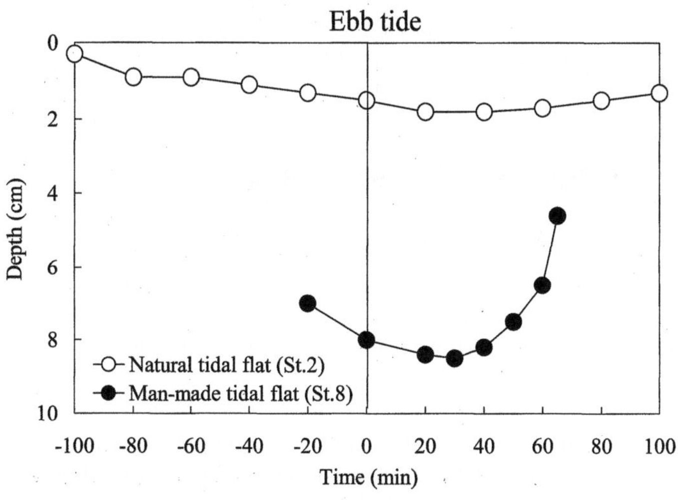

Fig. 1.4. Diurnal changes of groundwater levels in natural and man-made tidal flat

1.2.3 Biomass of bacteria and macrobenthos

Fig.1.5 shows seasonal changes of bacterial population at the surface of the natural (0-2 cm at St.2) and man-made (0-5 cm at St.8) tidal flats. Bacterial population in soils of the natural tidal flat ranged from 1.6 to 3.6 × 10^9 cells g-dry^{-1} but seasonal change was not observed. Similar result was obtained for bacterial population at the man-made tidal flat. Fig.1.6 shows bacterial populations at the surface of soil in 10 tidal flats. Bacterial populations ranged from 1.6 to 2.5 × 10^9 cells g-dry^{-1} in the natural tidal flats while the populations in the man-made tidal flats were in the range of 0.07 to 0.38 × 10^9 cells g-dry^{-1} except for St.10. The bacterial population in soil of the man-made tidal flat was one to two orders of magnitude lower than in soil of the natural one. However, the population at St.10 was 2.3 × 10^9 cells g-dry^{-1}. This was the highest among the samples collected from the man-made tidal flats and was similar to those in the natural tidal flats. Fig.1.7 shows vertical profiles of bacterial population in soils at St.2 and St.8. The bacterial populations did not change by depth at both tidal flats. Organic carbon supplied from seawater must be different between infiltration and non-infiltration zones that were from the surface to 2 cm in depth and below 2 cm at St.2, and the surface to 9 cm and below 9 cm at St.8, respectively. Results indicate that bacterial population is independent of organic carbon supplied.

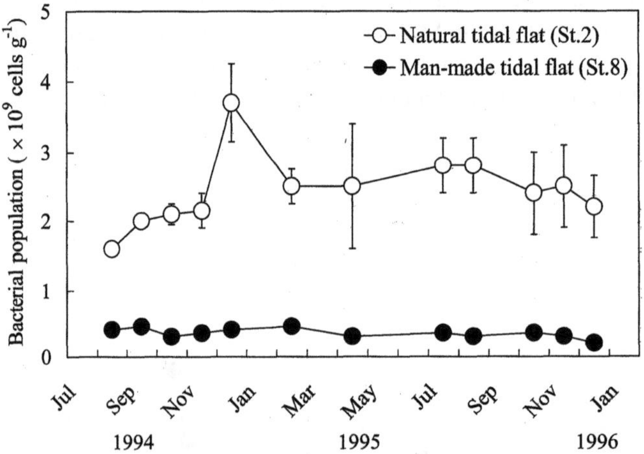

Fig. 1.5. Seasonal change of bacterial population in natural and man-made tidal flats

Fig.1.8 shows the biomass (as wet weight) and population of macrobenthos in the natural and man-made tidal flats. No remarkable difference between the natural and man-made ones was noted in the biomass. The percentages of bivalve to total biomass were extremely high at St.7 and St.8 due to artificial introduction of clam (principally *Ruditapes philippinarum*). Havens et al. (1995) reported that the biomass of macrobenthos in two natural tidal marshes and a man-made one were nearly the same and an apparent difference could not be found among them. The predominant species of macrobenthos in both tidal flats was Polychaeta. At St.10, however, Mollusca was the predominant species. Significant differences in both species and population of macrobenthos were not observed between the natural and the man-made tidal flats.

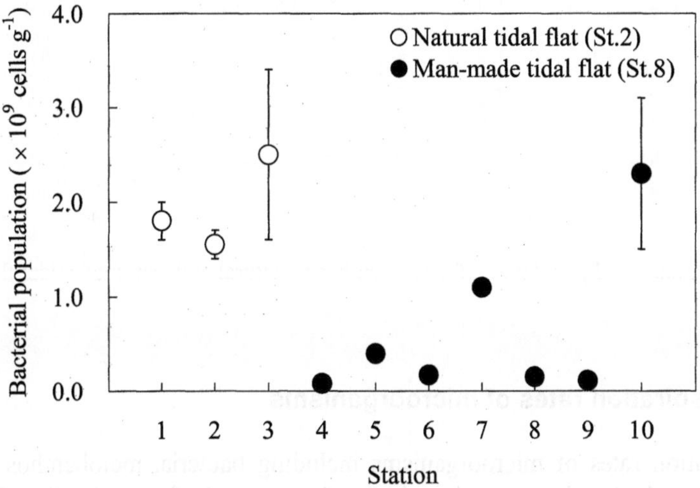

Fig. 1.6. Bacterial population in 10 tidal flats

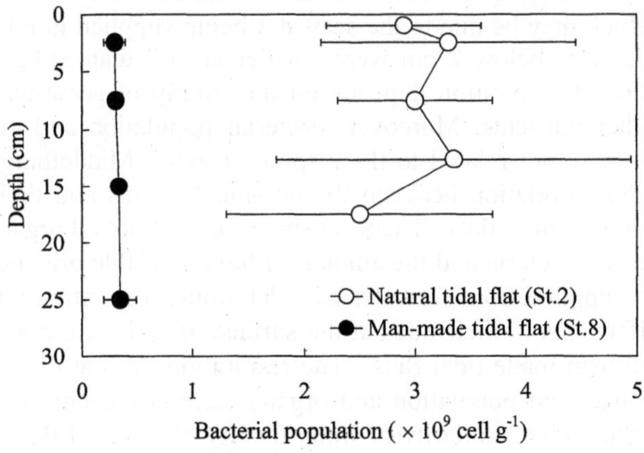

Fig. 1.7. Vertical profiles of bacterial population in soils of natural and man-made tidal flats

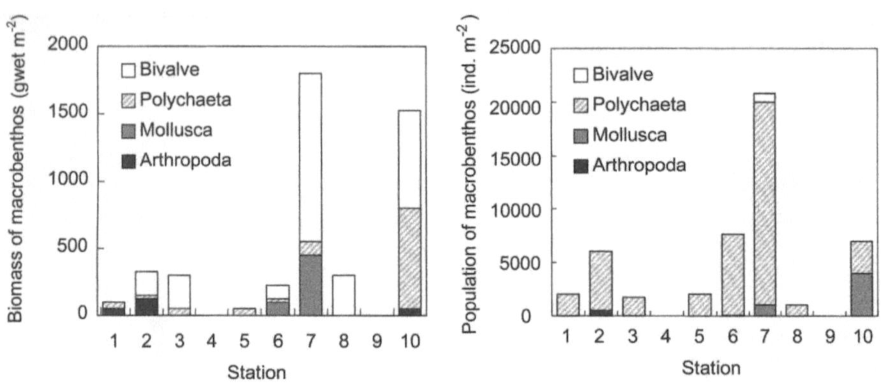

Fig. 1.8. Biomass and population of macrobenthos in natural and man-made tidal flats

1.2.4 Respiration rates of microorganisms

Respiration rates of microorganisms including bacteria, meiobenthos and attached phytoplankton were measured after the removal of macrobenthos from the soil sample. Fig.1.9 shows vertical profiles of microbial respiration rates in soil at St.2 and St.8. The respiration rates at the surface of soil in 10 tidal flats were summarized in Fig.1.10. The respiration rate at the surface of soil was extremely higher than other soil samples below 2 cm at St.2. On the other hand, the difference of the respiration rates between the surface and deeper zones at St.8 was relatively small which may be due to the seawater being supplied to all zones at St.8. The respiration rates below 2 cm were smaller at St.2 than St.8. These results mean that bacterial respiration depends on the supply of seawater with organic carbon and other nutrients. Moreover, bacterial population and organic carbon content in soil were not related to the respiration rate. Middelburg et al. (1996) also reported no correlation between the organic contents and their degradation rates in soils in river tidal flats. These results mean that not all organic contents in soil are utilized by bacteria and the amount of biodegradable organic matter existing in soil and supplied from seawater may determine respiration rate in soil. No significant differences in their rates at the surface of soil were observed between the natural and man-made tidal flats. The respiration rates at different tidal flats with different bacterial population and organic carbon content being almost the same may be due to the similarity of water quality of the tidal flats located in the same bay in terms of biodegradable organic matter content.

1.2.5 Discussion

Both physico-chemical and biological characteristics between natural and typical man-made tidal flats were significantly different on silt and organic carbon contents of soil, groundwater level and bacterial biomass. Vertical profiles of microbial respiration rates and ORP in soil of natural and man-made tidal flats were also different and related to groundwater level, that is, the zone where seawater is supplied keeps oxidative condition and high respiration activity of microorganisms due to organic and inorganic matters and oxygen supplied from seawater. Groundwater level in tidal flat may be determined by soil composition. Large amount of fine particles such as silt keeps water in soil structure and groundwater level does not fall significantly even ebb tide. Therefore, natural tidal flat with high silt content shows development of reducing zone to near the surface of soil and high respiration activity of microorganisms specifically at the surface of soil, whereas typical man-made tidal flat with low silt content shows oxidative condition until the deep zone and high respiration activity at all depths.

The bacterial population in soil of natural tidal flat showed one to two orders of magnitude higher than in soil of the typical man-made tidal flat even though microbial respiration rates were almost the same. Meyer-Reil et al. (1984) and DeFlaun et al. (1983) reported that bacterial population was affected by the size distribution of soil and bacterial population was high in soil at small size. Although organic matter supply from seawater is important to keep bacterial population, spatial habitat may be more important in these tidal flats located at Hiroshima bay.

Accumulation of fine particles such as silt in soil is proven to be very important in determining the physico-chemical and biological characteristics at tidal flat through this experimental approach. If fine particles can be accumulated in man-made tidal flat sediment, the man-made tidal flat may be a natural like tidal flat. The existence of St.10, which has almost the same characteristics as natural tidal flat, show man-made tidal flat can be made like natural one.

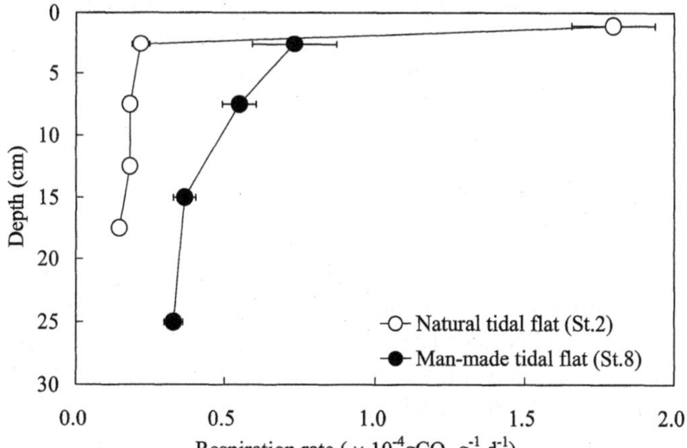

Fig. 1.9. Vertical profiles of microbial respiration rates in soil of natural and man-made tidal flats

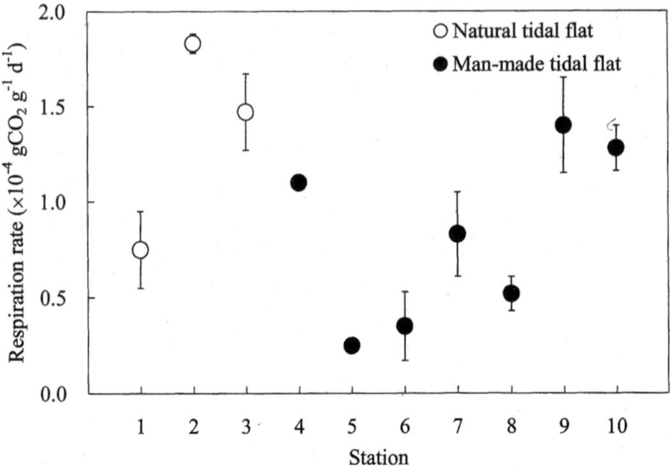

Fig. 1.10. Microbial respiration rates at the surface of soils in 10 tidal flats

1.3 Key factors in determining a tidal flat characteristic

1.3.1 Transfer experiments

The soil in a man-made tidal flat (St.8) was transferred to a natural one (St.2) and vice versa was conducted to find out controlling factors of tidal flat ecosystems. These two tidal flats had almost the same area (47,000 m^2 at St.2 and 50,000 m^2 at St.8). The site transferred was + 1.2 m datum level. The portions of the soil surface to 10 cm in depth were collected from each tidal flat for transfer and macrobenthos in the soil was removed. The soil in the transfer site was previously removed in the area of 60 cm × 60 cm × 10 cm in depth. Before the soil was transferred, nylon mesh with 1 mm in pore size was used to cover the surface of the hole to prevent invasion of macrobenthos.

Silt content, bacterial population and ignition loss were traced after the transfer of each soil. Fig.1.11(a) shows variation in each parameter in soil of the man-made tidal flat transferred into the natural one. All parameters increased gradually after transfer and reached steady state at day 44. The silt content changed from 0.4% at start to 2.9% at day 44, which was the comparable value as the silt content in soil of the natural tidal flat nearby shown in Table 1.2. Bacterial population and ignition loss also indicated almost the same values as those in soil of the natural one. Fig.1.11(b) shows variation in each parameter in soil of the natural tidal flat transferring to the man-made one. The silt content rapidly decreased after transfer into the man-made one and reached less than 1% within 2 weeks. Both bacterial population and ignition loss indicated rapid decrease as well. As a result, three parameters in the transferred soil of the natural tidal flat reached almost the same values as those of the man-made tidal flat nearby. These results suggest that accumulation or loss of silt and organic matter expressed by ignition loss in soil easily changes with environmental condition on site. It is difficult to think that organic matter and nutrient supply from seawater for bacteria are different in both tidal flats since the water quality in offshore water in both tidal flats are almost the same (Maritime Safety Agency 1997). Furthermore, the change in bacterial population along with the changes of silt and ignition loss strongly indicates that bacterial population is related to silt and organic matters content of soil.

Fig.1.12 shows vertical profiles of ORP of soils transferred between the natural and man-made tidal flats at day 78. ORP at all layers of the original soil in the man-made tidal flat showed at positive values, whereas, in the natural one, ORP showed negative values except the surface. However, the transferred soil from man-made tidal flat showed the same vertical profile of ORP as the soil in the natural one after day 78 and the transferred soil from the natural one showed the same vertical profile of ORP as the soil in the man-made one. The change in ORP by transfer might have been induced by accumulation or loss of silt and organic matter, which are related to infiltration of seawater with oxygen into sediment.

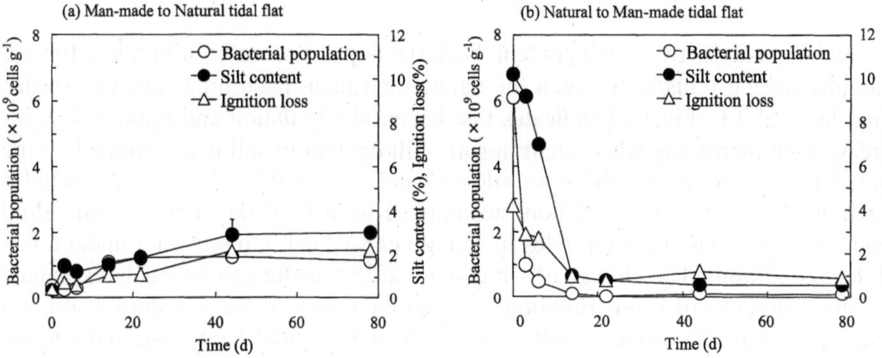

Fig. 1.11. Variation in bacterial population, silt content and ignition loss in (a) soil of man-made tidal flat transferred into natural one and (b) soil of natural tidal flat transferred into man-made one

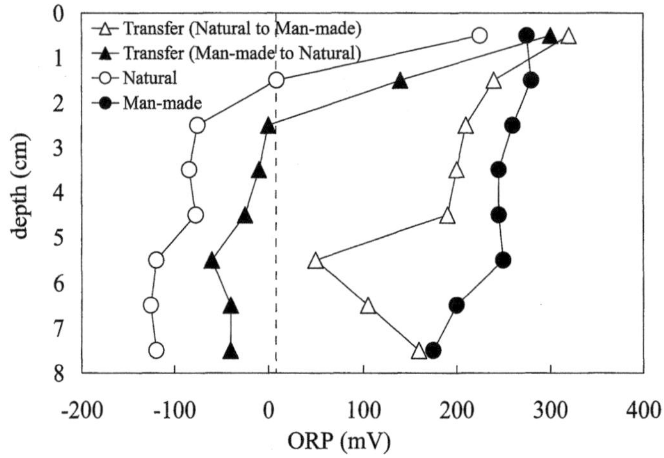

Fig. 1.12. Vertical profiles of oxidation-reduction potential of soils transferred between natural and man-made tidal flats

1.3.2 Relationship between silt content and bacterial population

Relationship between silt content, bacterial population and ignition loss for soil samples during transfer between the natural and man-made tidal flats was evaluated in Fig.1.13. Fig.1.13 indicates that bacterial population and ignition loss increase with increasing silt content in soil. Silt content in soil is determined by the balance of inflow and outflow of silt into and from a tidal flat. Small particles such as silt carried from offshore accumulate on a tidal flat under a calm fluid condition, whereas small particles in soil are carried off into seawater under a turbulent condition. The change of silt content after transfer can be easily explained by the difference of fluid condition between the origin of the transferred soil and the destination. For instance, silt in soil of the natural tidal flat transferred into the man-made one was carried off into seawater due to the strong fluid of seawater on the man-made one. The organic particles such as detritus may accumulate on or be carried off from a tidal flat by the same phenomenon as silt.

However, change of bacterial population in soil cannot be explained by the same phenomenon since the specific gravity of bacteria is almost the same as seawater. Increase of surface area with the increase of silt content in soil, which can cause an increase of bacterial attachment place, or increase of organic matters, which is food for bacterium, may result to the increase of bacterial population.

Transfer experiment suggested that bacterial population is related with silt content in soil. The relationship was confirmed in a simplified tidal flat system (Fig.1.14), which only contained organic matter free soil of various amounts of silt and seawater without suspended matter. Bacteria were removed from the soil before usage to unify the experimental condition on each soil. The seawater with bacteria was induced to the tidal flat system. The equipment was operated with a tidal fluctuation of 12 h cycle at 20 °C in a dark condition. The highest water level reached the top of the column and the lowest water level reached the bottom of the column.

Bacterial population became stable after 40 to 55days of operation. The population showed a strong relationship with the silt content in soil (r=0.99) as shown in Fig.1.15. Meyer-Reil et al. (1978) and Weise and Rheinhe-imea et al. (1978) reported that most of the bacteria in a tidal flat and sandy beach attached on the surface of soil. Meyer-Reil et al. (1984) and DeFlaun et al. (1983) also reported that bacterial population was affected by the size distribution of soil and as an effect, bacterial population was high in soil with small size. The increase of silt content in soil implies an increase in the surface area of soil per unit weight or volume. Effect of organic matter on bacterial population in soil can be neglected in this experiment since organic matter free soil and seawater without suspended matter were used. Therefore, bacterial population in soil depended on silt content.

Glucose was added to seawater to clarify the effect of food on bacterial population. Bacterial population also showed a strong relationship with the silt content (r=0.99) under rich food condition as shown in Fig.1.15. However, bacterial population with glucose addition showed one to three orders of magnitude higher than that without glucose addition.

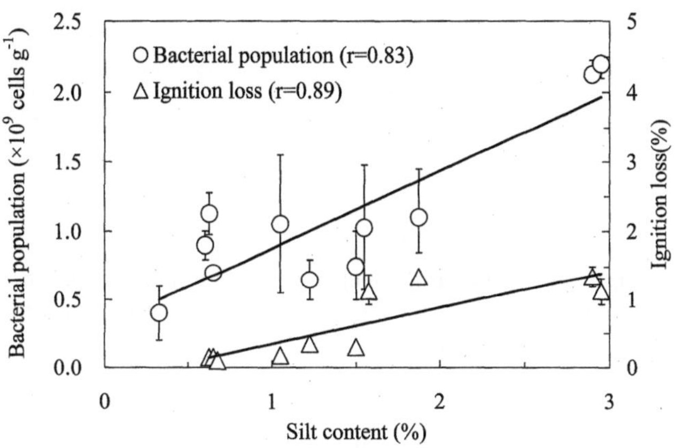

Fig. 1.13. Relationship between bacterial population, ignition loss and silt content for soil samples during transfer between natural and man-made tidal flats

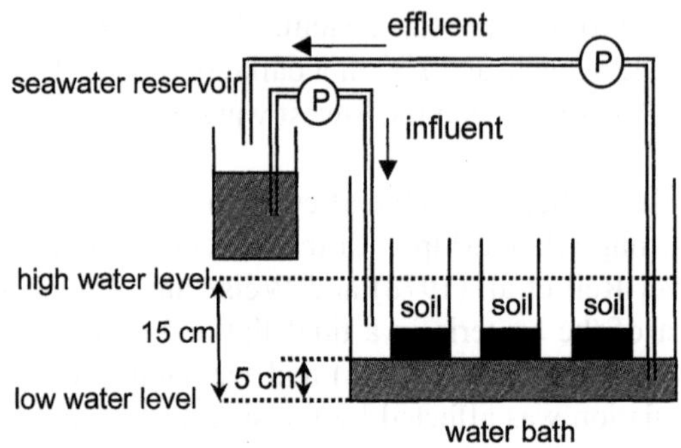

Fig. 1.14. Schematic diagram of experimental setup

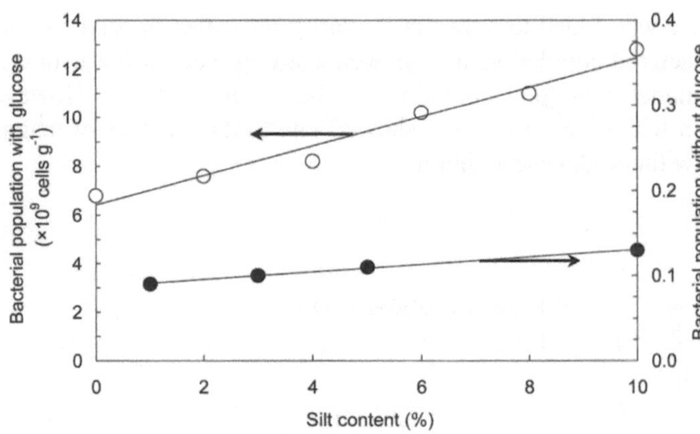

Fig. 1.15. Relationship between bacterial population, ignition loss and silt content in soil samples taken at the end of experiments with or without the addition of glucose

High amount of ignition loss in soil of a tidal flat does not directly mean that high amount of food is supplied for the bacteria in the tidal flat. Most of the organic matter for bacteria must have been supplied from seawater and ignition loss just shows that organic matter accumulated in soil which contains much refractory organic matter to bacteria. However, high amount of ignition loss in soil may mean that high amount of organic particle deposits in the tidal flat which may increase food for bacteria. Therefore, bacterial population in soil depends on surface area of soil for habitat and food. Food for bacteria may become rich by deposition of organic particles in tidal flat under calm fluid condition compared with that under strong fluid condition even though the quality of seawater flowing in the tidal flat is the same.

1.4 Fluid flow on a tidal flat

1.4.1 Outline of study sites

The study sites were selected at St.9 and St.10 in Fig.1.1. Both sites are man-made tidal flats bordering each other. Therefore, fluid condition and water quality of offshore are almost the same. However, the soil structure was quite different between two tidal flats as shown in Table 1.1 and 1.2. Bacterial populations at St.9 and St.10 were 0.11 and 2.32 × 10^9 cells g^{-1}, respectively. St.9 was typical man-made tidal flat, whereas St.10 was natural-like man-made tidal flat, that is, bacterial population, organic matter content and silt content of soil in St.10 were much higher than those in St.9.

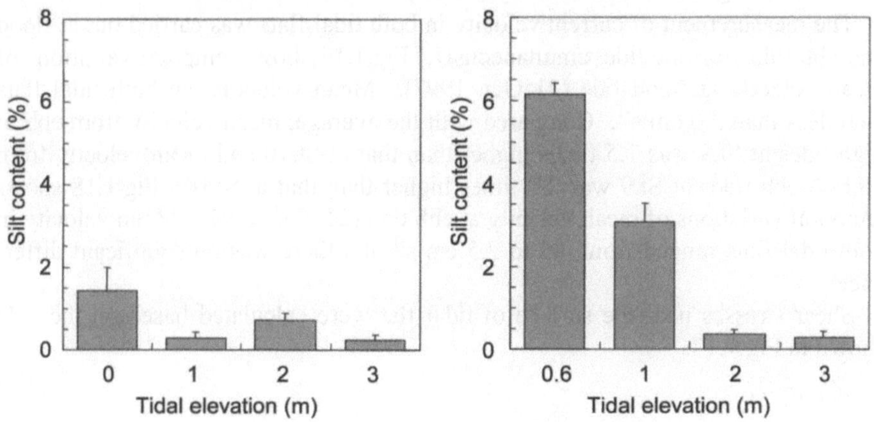

Fig. 1.16. Accumulation of silt particles at different tidal elevation at St.9 (right) and St.10 (left)

Fig.1.16 shows the accumulation of silt particles at different tidal elevation. Silt contents at + 3 m from the datum level was low and almost similar for both tidal flats. It is most likely that both tidal flats were constructed with soil of low silt content because it has little time to be flooded at + 3 m. However, silt contents at + 0.6 m and + 1.0 m in St.10 were very high indicating the accumulation of silt after construction. The only possible source of silt for St.9 and St.10 is seawater because river water does not flow into these tidal flats. The qualities of seawater at both stations are considered to be the same because of their proximity.

1.4.2 Shear stress

Fluid condition was analyzed in both tidal flats to clarify the difference of soil structure. Current velocity in overlaying seawater was measured at 5 cm from soil surface and +1.5 m. Velocity was measured at 10 Hz of frequency using acoustic Doppler current meter (ADV-field, NORTEK, INC., Norway). Two horizontal (u and v) and one vertical (w) components of velocity were obtained; u is the velocity from land to sea and v is an orthogonal velocity to u. Average horizontal velocity ($\overline{U}$) was determined from the average of each horizontal velocity for 10 min:

$$\overline{U} = \overline{\sqrt{u^2 + v^2}} \tag{1.1}$$

Where:

u, v: horizontal current velocity (cm s^{-1})

$\overline{U}$: average horizontal velocity (cm s^{-1})

The measurement of current velocity in both tidal flats was carried out at flood and ebb tides for one tide simultaneously. Fig.1.17 shows temporal variations of mean velocity at flood tide (21 Oct. 1997). Mean velocities in both tidal flats were less than 3.0 cm s^{-1}. Compared with the average, mean velocity from ebb to high tides at St.9 was 1.5 times higher than that at St.10 and mean velocity from high to ebb tides at St.9 was 2.0 times higher than that at St.10. Fig.1.18 shows temporal variations of mean velocity at ebb tide (21 Nor. 1997). Mean velocity in both tidal flats ranged from 0.5 to 2.5 cm s^{-1} and there was no significant difference.

Shear stresses near the surface of tidal flat were calculated based on the data shown in Fig.1.17.

$$u' = u - U \tag{1.2}$$

$$v' = v - V \tag{1.3}$$

$$w' = w - W \tag{1.4}$$

$$\tau_{uw} = -\rho u'w' \tag{1.5}$$

$$\tau_{vw} = -\rho v'w' \tag{1.6}$$

$$\tau_{total} = \sqrt{\tau_{uw}^2 + \tau_{vw}^2} \tag{1.7}$$

Where:

u', v', w' : fluctuation current velocities (cm s^{-1})

U, V, W : average velocities (cm s^{-1})

τ_{uw}, τ_{vw} : 2-dimension shear stresses (N m^{-2})

τ_{total} : combined shear stress of τ_{uw} with τ_{vw} (N m^{-2})

ρ : density of seawater (kg m^{-3})

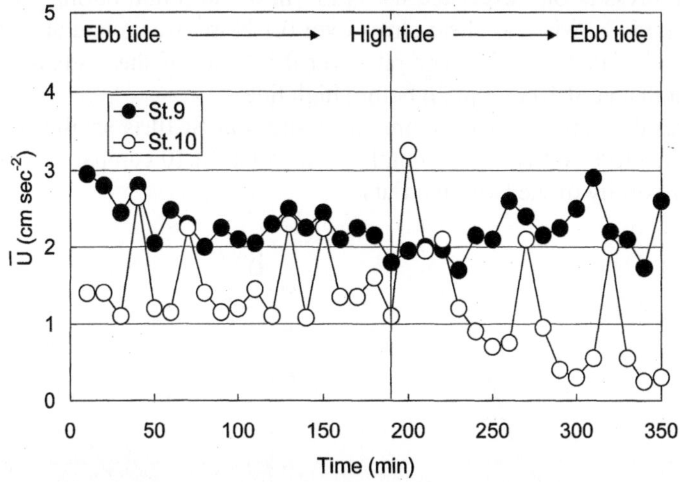

Fig. 1.17. Temporal variations of mean velocity at flood tide (30 Oct. 1997)

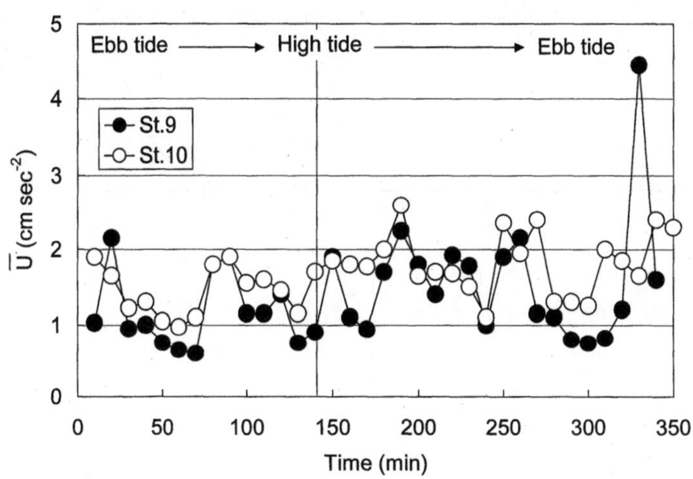

Fig. 1.18. Temporal variations of mean velocity at ebb tide (21 Nor. 1997)

Cumulative distributions of shearing stress for 10 min at high tide, 1h and 2h after high tide and 1h and 2h before high tide are shown in Fig.1.19. Although shear stress at high tide was the same between two stations, shear stress at St.9 was higher than that at St.10 at another calculated periods. The critical shear stress for suspension of silt particles was calculated as 0.1 N m^{-2} based on the experimental data by Mehta *et al.* (1982) using equation proposed by Iwagaki (1985). The rate over 0.1 N m^{-2} of shear stress was small at both tidal flats. However, shear stress at St.9 exceeded 0.1 N m^{-2} in 0.5 to 3 min during 10 min in all calculated periods, whereas shear stress over 0.1 N m^{-2} was found at only two periods in 5 calculated periods. The rate over 0.1 N m^{-2} of shear stress at St.9 was higher than that at St.10 except 2h before high tide.

These results indicate that bottom shear stress at St.10 is smaller than that at St.9 and, therefore, silt easily accumulates in soil at St.10 compared with St.9 despite the proximity of the two tidal flats.

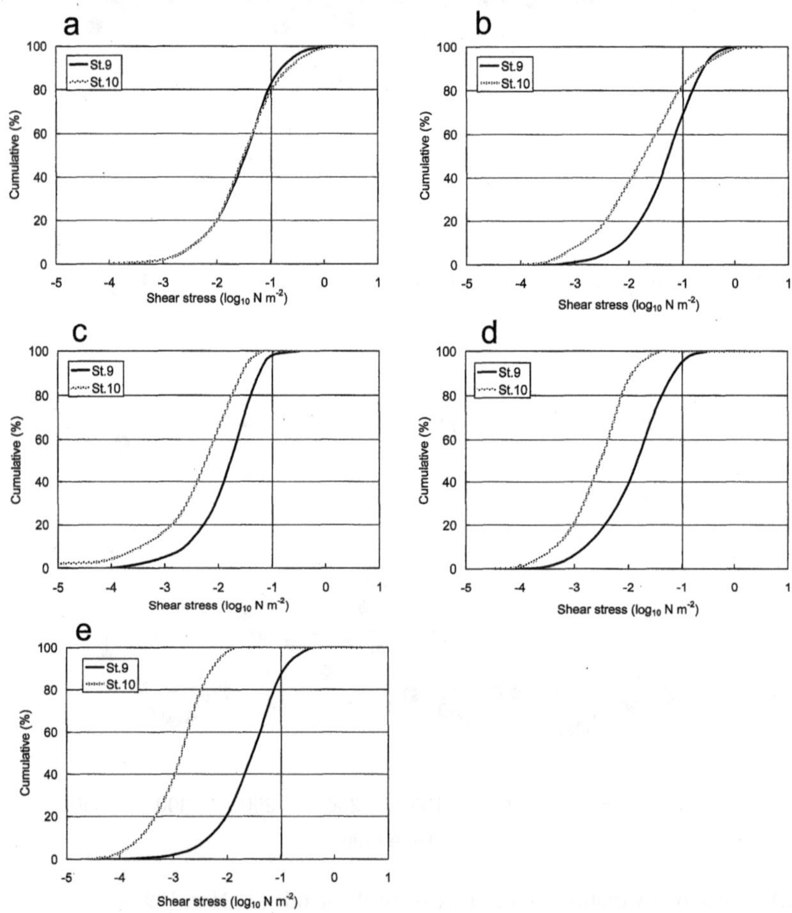

Fig. 1.19. Temporal variation of shear stress at 30 Oct. 1997, (a) before 2 h from high tide, (b) before 1 h from high tide, (c) high tide, (d) after 1 h from high tide, (e) after 2 h from high tide

1.4.3 Sedimentation of sand

To clarify the difference of resuspension of sand between St.9 and St.10, a sediment trap was set at + 1.2 m of elevation for one tide. The amount and particle size distribution of resuspended and trapped sediment at both tidal flats are shown in Fig.1.20. The amount of trapped sediment at St.9 was 133 g m^{-2} and was 2.3 times larger than that at St.10 (59 g m^{-2}). Furthermore, the particle distributions of trapped sediment between St.9 and St.10 were quite different and relatively big particles were trapped at St.9. Over 1000 μm of particles were not trapped at St.10, whereas about 50 g m^{-2} was trapped at St.9. Large amount of silt (less than 75 μm) trapped at St.10 indicates that silt undergoes sedimentation and resuspension in the tidal flat.

Since large particles were trapped only at St.9 and the amount of trapped sediment was larger than that at St.10, the bottom shear stress at St.9 is larger than that at St.10. This conclusion from sediment trap investigation is coincident with the conclusion from shear stress analysis. The difference of bottom shear stress between the two tidal flats results to differences in not only physico-chemical but biological characteristics because biological structures are strongly dependent on silt content in soil as mentioned in 1.3.2.

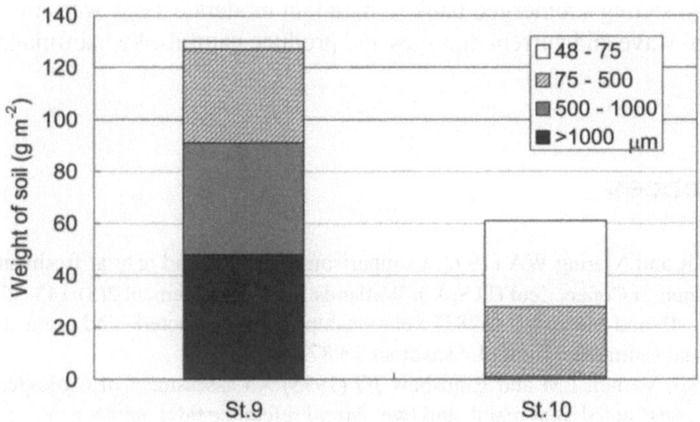

Fig. 1.20. Particle size distribution of trapped sediment at St.9 and St.10

Why is fluid condition different between the two tidal flats in proximity? The differences of construction characteristics between two man-made tidal flats are slope and existence of submerged bank. The slope of St.10 was 0.7/100 and was quite smaller than that of St.9 (4/100), as shown in Table 1.1. The slope of St.10 was smaller than those of natural tidal flats (1/100 to 1.6/100) as shown in Table 1.1. At St.10, submerged bank has been constructed to prevent erosion by current and wave and to maintain moderate slope. Submerged bank and moderate slope reduce the wave and current energies into tidal flat. Hence, bottom shear stress in tidal flat is reduced and silt was accumulated in soil at St.10.

1.4.4 Discussion

Most man-made tidal flats studied showed different physico-chemical and biological characteristics from natural ones. Low silt and organic carbon contents, low groundwater level, oxidative soil condition and low bacterial biomass were observed in man-made tidal flat. The different characteristics were explained by the difference of accumulation of silt and organic matter between man-made and natural tidal flats indicating fluid conditions of both tidal flats were quite different. Man-made tidal flat is constructed at a shoreline where natural tidal flat is not developed indicating that fluid condition in the site is not suitable for tidal flat development primarily due to the strong wave and/or current. Even if tidal flat is constructed using the same soil with large amount of fine soil as natural tidal flat, fine soil may be carried off into seawater and the tidal flat consequently becomes coarse grained sandy beach. Therefore, controlling the fluid condition in overlying seawater in tidal flat becomes very important to make natural-like man-made tidal flat. The investigation of natural-like man-made tidal flat, St.10, gave relevant information to control fluid condition. Constructing tidal flat with moderate slope and setting submerged bank to maintain moderate slope were the key factors to reduce wave and current energies and produce natural-like man-made tidal flat.

References

Confer SR and Niering WA (1992) Comparison of created and natural freshwater emergent wetland in Connecticut (U.S.A.). Wetlands Ecol. Management 2(3):143-156

DeFlaun MF and Mayer LL (1983) Relationships between bacteria and grain surfaces in intertidal sediments. Limnol. Oceanogr 28:873-881

Havens KJ, Varnell LM and Bradshaw JG (1995) An assessment of ecological conditions in a constructed tidal marsh and two natural reference tidal marshes in coastal Virginia. Ecol Eng 4:117-141

Iwagaki Y (1956) Hydrodynamical study on critical tractive force. Transactions of the Japan Society of Civil Engineers 41:1-21

Kimura K (1994) The function of water purification in constructed tidal flat. Jpn Bottom Sediment Management Assoc 60:50-81

Lee JG, Nishijima W, Mukai T, Takimoto T, Seiki T, Hiraoka K and Okada M (1997) Comparison for structure and functions of organic matter degradation at natural and constructed tidal flats - A case study in Hiroshima bay -. Journal of Japan Society on Water Environment 20:175-184

Lee JG, Nishijima W, Mukai T, Takimoto T, Seiki T, Hiraoka K and Okada M (1998a) Quantification of purification ability for organic matter at natural and constructed tidal flats and the role for purification in Hiroshima bay. Journal of Japan Society on Water Environment 21:149-156

Lee JG, Nishijima W, Mukai T, Takimoto T, Seiki T, Hiraoka K and Okada M (1998b) Factors to determine the functions and structures in natural and constructed tidal flats. Water Research 32:2601-2606

Lee JG, Nishijima W and Okada M (1999a) Study on selection of location in constructed tidal flat and controlling factors on soil structure. Journal of Japan Society on Water Environment 22:595-599

Lee JG, Cheong CJ, Nishijima W, Okada M and Baba E (1999b) Effect of seawater flow on the soil structure in tidal flat. Journal of Japan Society on Water Environment 22:677-682

Maritime Safe Agency (1997) Ocean environment in Hiroshima Bay; Part 5 of Report. Hydrographic Department, Japan

Methta A and Partheniades E (1982) Resuspension of deposited cohesive sediment beds. Coastal Engineering 18:1569-1588

Meyer-Reil LA, Dawson R, liebezeit G and Tiedge H (1978) Fluctuations and interactions of bacterial activity in sandy beach sediments and overlying waters. J Mar Biol. 48:161-171

Meyer-Reil LA (1984) Bacterial biomass and heterotrophic activity in sediments and overlying waters, in Heterotrophic Activity in the sea. Edt. By Hobbie JE and Williams PJLeB. Plenum Press, New York, pp523-546

Middelburg JJ, Klaver G, Nieuwenhuize J, Wielemaker A, Vlug T and van der Nat J.F. WA (1996) Organic matter mineralization in intertidal sediments along an estuarine gradient. Mar. Ecol Prog Ser 132:157-168

Miyoshi K, Shimati T and Kimura K (1990) A capacity of purification at constructed tidal flat. Tokyo Metropolitan Research Institute of Environment Science (An annual report), pp120-125

Ogura O and Imamura H (1995) Creative technique for constructed tidal flat. Jpn Bottom Sediment Management Assoc 64:61-78

Weise W. and Rheinheimer G (1978) Scanning electron microscopy and epifluorescence investigation of bacterial colonization of marine sand sediments. Microb Ecol 4:175-188

2 Effects of waves and tide on tidal flat ecosystems

2.1 Introduction

Coming wave to coastal area has an essential effect on ecosystems. In shallow water region such as sandy beach and tidal flats, wave run-ups due to breaking waves, whose height are around 5 cm to 10 cm, are often observed at the slope. In addition to the tidal motion, these small waves may affect water flow inside the seabed. Internal flow in the seabed has close relation with biological activities in these areas. Seawater is a kind of transport medium for oxygen and nutrients. As stated in Chap. 1, the number of bacteria inside the seabed is known to closely correlate with the silt content because the presence of such small particles increases the wet surface area for bacteria habitation. The small particles easily move along with the flow. Therefore, to determine the internal flow characteristics of tidal flat is important for understanding of its ecological role. Studies on seawater transport in the surf zone have been carried out mostly for large waves of several meters: (Riedel 1971; McLachlan 1982). However, studies for small waves of several cm high are scarce, hitherto.

This chapter deals with the results of studies by use of an experimental facility to obtain better understanding of internal flow inside sandy beach and tidal flats. This facility aims to simulate a small segment of a tidal flat in a laboratory size. It is equipped with a wave maker, tide controlling device, temperature controller and computer controlling system. At first, studied for sandy beach were made. For visualization of internal flows, transparent glass beads were used instead of real sediments of tidal flats. Observations on seawater infiltration by breaking waves, and movement of silts were carried out. It became possible to quantify the water volume pumped by breaking waves. It was found that for relatively steeper seabed, silt particles are sucked out from seabed at the point of wave breaking.

In the next stage, experiments by use of real sediments taken from tidal flats were made to know the effects of waves and tide on the ecosystems in tidal flats. This experiment showed that the ecosystem of tidal flat was changed with and without waves. Without waves a different ecosystem appears. It is understood that waves are indispensable for sound development of tidal flat where macrobenthic community live as the bases of food chain in seabed.

2.2 Method of experiments on internal flow in seabed

2.2.1 Flow visualization inside sandy beach

In order to visualize internal flow in sandy beach, transparent glass beads were used. The glass beads of 1 mm diameter were chosen based on the measured mean particle size of man-made tidal flats, Stations 9 and 10 at Nohmishima of Etajima Bay in Hiroshima: (Lee et al. 1999).

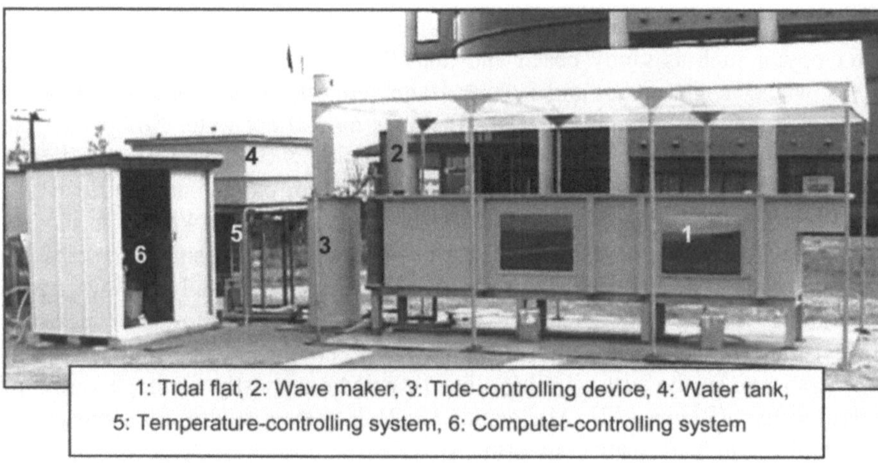

Fig.2.1. Tidal flat experimental facility

Fig. 2.1 shows the tidal flat experimental facility (JAPAN AQUA TEC Co., Ltd. JBES-2500A), which is composed of a tidal flat (glass beads), wave maker, tide controlling device, temperature controller and computer controlling system. The facility made of fiber-reinforced plastics (FRP) textile has windows for observation and has a dimension of 5 m long, 0.8 m wide and 1 m high. It is filled with transparent glass beads of 2.5 specific weight under normal atmospheric condition. The slope of the sandy beach is manually adjusted. A plunger-type wave maker generates waves with a maximum height of 50mm at periods from 0.3 to 3.0 s. The tide is controlled at periods between 1 to 7 h. Synthetic seawater was produced with a 3.5% salinity using tap water and commercial sea salt (AQUA-TEC. Co., Sealife). Water temperature was controlled at 15 ± 1℃. To visualize infiltration of seawater into the seabed, potassium permanganate ($KMnO_4$) of 0.02 mol/L concentration was used as a tracer which has a comparable specific weight with seawater. For observation of silt movement, black anthracite with a density of

1.5 and size of less than 75μm was used instead of actual silt (sand particles with a mean diameter of less than 75μm).

2.2.2 Wave and tide control

In an actual tidal flat, wave and tide actions are combined. However, under laboratory conditions, wave action alone can be treated separately, because the periods of both phenomena are completely different. That is, the period of waves is on the order of a few seconds. On the other hand, the period of tide is on the order of 12 h. Both phenomena, therefore, could be treated separately.

The experimental facility was designed so as to observe the seawater movement by breaking waves at the surf zone. In this study low-energy waves often observed in tidal flats were considered, where the breaking wave heights are around 50mm and the periods are approximately 1 to 2 s. Breaking wave generated in this experiment are classified as the collapsing type: (Sorensen 1997). The vertical translation velocity of seawater by tide was determined to be 0.009 cm/s based on the mean tidal range of 2 m for 6 h in Hiroshima Bay. Therefore, in the present facility the water level due to the tide was controlled at a velocity of 0.009 cm/s in a tidal range of 20 cm. In the present experiments, observations of seawater infiltration were made in the following three cases: wave action alone, that of tide alone and the combination of wave and tide actions.

2.2.3 Quantification of seawater infiltration in sandy beach

In the case of action of wave alone it was found that seawater infiltration due to breaking waves in a sloping sandy beach was very simple as shown in Fig.2.2. The colored dye (potassium permanganate) infiltrates progressively into the sandy beach in a semicircular pattern with the period of coming waves. This means that the flow spreads radially. That is, fresh seawater infiltrates radially into the sandy beach from a point near the run-up point of breaking waves. This flow phenomenon is similar to that observed from the source, which is often used in hydrodynamics. Based on this finding it became possible to quantify the amount of seawater infiltration based on the wave action. That is, from the video recording of the cross section of colored dye infiltration, seawater infiltration can be determined by integration of the area.

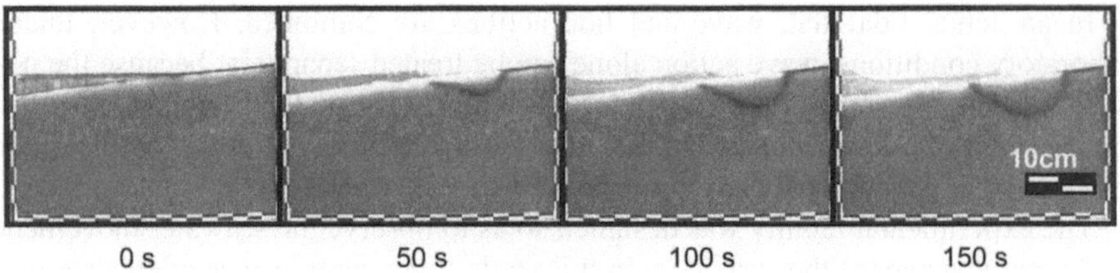

Fig. 2.2. Infiltration behavior of seawater due to wave action

2.3 Role of wave and tide on seawater infiltration in sandy beach

2.3.1 Fresh seawater infiltration

Figure 2.3 shows infiltration rates of seawater determined from video recording (SONY Digital Handy Camera DRC-VX1000) based on wave action alone for different slopes of seabed. In this experiment, the breaking wave height is 50 mm, at a period of 1 s, and glass beads of 1 mm diameter. It is shown that the speed of propagation of the area infiltrated by the colored dye was approximately 2 cm^2/s in the case of a 10/100 slope. Assuming that fresh seawater infiltrates radially from one point, it is estimated that the velocity of seawater infiltration is about 3.2 mm/s at points below 20 mm from the seabed surface. Furthermore, it is estimated that the wet volume due to infiltration is about 4.3 m^3 per 1-m width of tidal flat during one tide cycle from low to high for 6 h.

Figure 2.4 shows changes in seawater levels due to tidal motion alone during the first 4000 s (about one hour). In this case, seawater level simply changes vertically with the specified velocity 0.009 cm/s, which corresponds to a 2-m tide for 6 h. Figure 2.5 shows infiltration behavior of fresh seawater under the combination of wave and tide actions during the first 400 s. At the crossing point of water level and sloping seabed, fresh seawater infiltrates in a semicircular pattern as observed in the case of wave action alone. There are some differences in colored dye movements for rising and falling tides. On the whole, however, except at the crossing point of wave and seabed, as time increases on the order of hours, change in water level behaves in a manner similar to that of tide action alone.

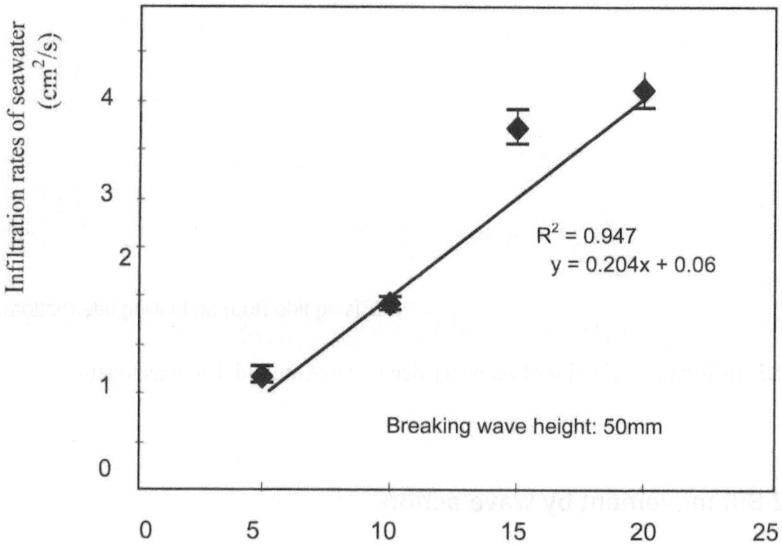

Fig. 2.3. Infiltration rates of seawater at different seabed slope

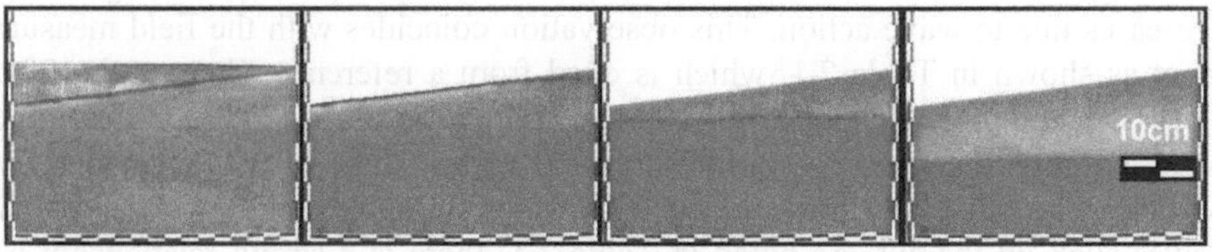

Fig. 2.4. Infiltration behavior of seawater due to tidal fluctuation

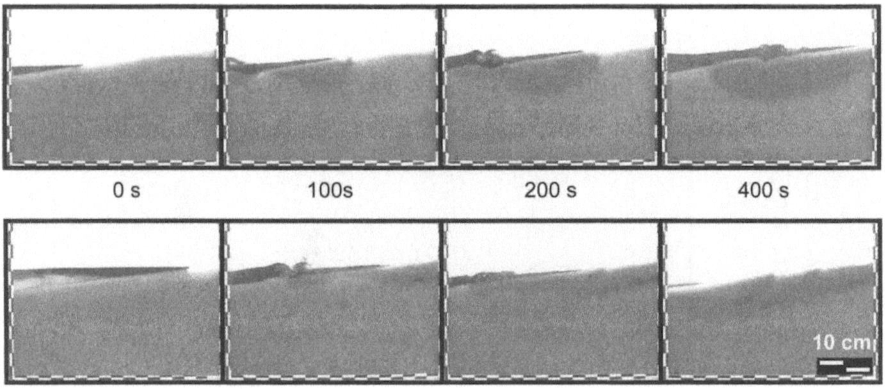

Fig. 2.5. Infiltration behavior of seawater due to wave and tidal actions together

2.3.2 Silt movement by wave action

Figure 2.6 shows the behavior of silt under wave action alone. The silt is replaced with anthracite as mentioned previously and initially spread over the surface of seabed. Then waves were formed. A comparison was made for two cases, one for a lower slope 2.5/100 (top) and the other for a steeper slope 10/100 (bottom). At the lower slope, most of the anthracite remains near the surface of the seabed with increase of time. On the other hand, for the steeper slope, the anthracite penetrates deeper near the run-up point of breaking waves in a manner similar to that of fresh seawater.

However, as shown in Figure 2.7, which is a close-up picture of the steeper slope at 100 s, the silt is sucked out from the seabed at the point of wave breaking. This suggests that for the steeper slope in the seabed, the quantity of silt gradually decreases due to wave action. This observation coincides with the field measurement as shown in Table 2.1, which is cited from a reference: (Lee et al. 1999). That is, Station 9 has a slope of 4/100 and content of silt of 0.05%. On the other hand, the silt content is 3.42% in Station 10 with a slope of 0.7/100. Since fine particles like silt easily move along with water flow, upward flows at breaking waves suck up the silt. As noted in Table 2.1, there is a strong relationship between the silt content and the number of bacteria. That is, the steeper slope with a low content of silt has a lower number of bacteria.

In this experiment it was noted that breaking waves have two functions, one is to pump fresh seawater into the seabed near the run-up point, and the other is to suck seawater from the seabed at the point of wave breaking. These observations are supported well by numerical calculations in the following section. It is clear therefore that inside the seabed there is water circulation between the run-up point and wave breaking point. This phenomenon was visually demonstrated already by Longuet-Higgins: (Longuet-Higgins 1983). In the beginning of this study, the authors noted only one aspect of the water flow phenomenon, that is, the pumping effect of breaking waves near the run-up point. In 2000 Doi et al made clear that the inside water flow near the run-up point is a beginning part of the water circulation which was observed by Longuet-Higgins: (Doi et al. 2000). Thus the authors noted the importance of sucking function of breaking waves, because it is related to resuspension of small particles such as silt.

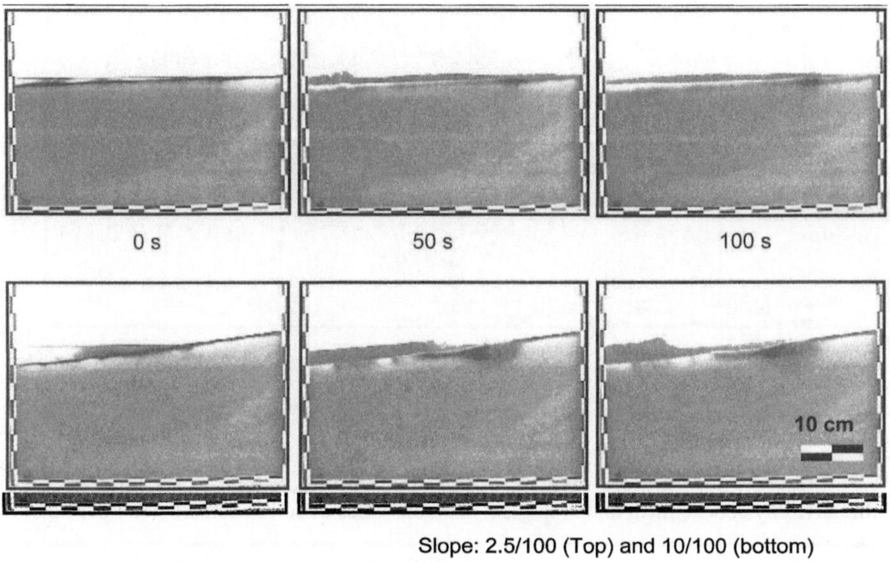

Fig. 2.6. Comparison of flow patterns of particulate matters through porous medium

2.4 Role of wave and tide on ecosystems in tidal flats

2.4.1 Measurement of seawater infiltration in tidal flat

Tidal flat ecosystems are important habitats for many commercial fishes and migratory birds. Macrobenthos in tidal flats are also an important food sources for many carnivores such as fishes and birds. In shallow coastal waters, environmental factor such as bottom topography, sediment characteristics and hydrody-

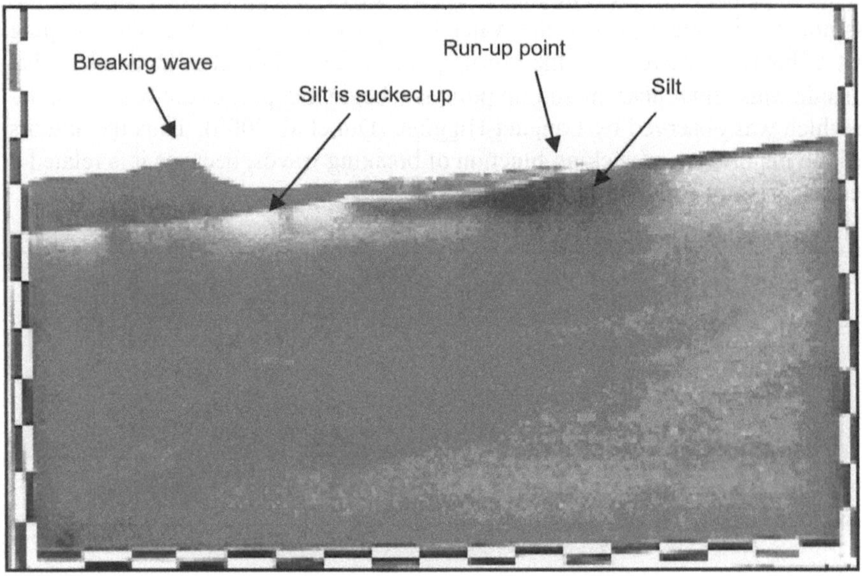

Fig. 2.7. A close-up picture of a steeper slope, shown in Fig. 2.6, at 100 s

Table 2.1. Physicochemical and biological characteristics of soil in constructed tidal flats

	Station 9	Station 10
Bacteria ($\times 10^9$ cells / g)	0.11	2.32
Organic carbon (mg C /g)	1.09	3.78
Silt content (%)	0.05	3.24
Median size (mm)	0.78	0.41
Year constructed	1987	1987
Slope	4 / 100	0.7 / 10

namic processes play a definite role on the structure of benthic communities through a variety of mechanisms. These seawater infiltrations supply both dissolved oxygen and particulate organic matters to benthic organisms. Therefore, waves and tides seem to be crucial factors for tidal flat ecosystems. The purpose of study is to clarify the effects of wave actions on benthic organisms in tidal flat ecosystems.

Seawater infiltration by wave actions

In the previous sections, glass beads were used instead of actual tidal flat sediments for visualization of internal flow of sandy beach. As the next step, for the study of ecosystems of tidal flat, actual tidal flat sediments were put into the experimental facilities. Fig.2.8 shows the grain size distribution of sediments used in this study. It is a characteristic that small particles less than 1 mm are contained largely.

Since actual tidal flat sediments are not transparent, the flow visualization method used for sandy beach is not able to use. Therefore, another method was developed here. Lithium nitrate ($LiNO_3$) was used as a conservative tracer to track the infiltration of seawater into the sediments by wave action, because lithium is not absorbed (Wrenn 1977). The lithium nitrate was dissolved in seawater of the tank to give an initial concentration of 10 mg/L under the constant water level. And then waves were applied for 3 h without any tidal movement. At the end of wave actions, sediment was sampled with a cylindrical acryl core split lengthwise and taped together (inner diameter 2.5 cm, length 30 cm) from 5 points at intervals of 7.5 cm between columns around the breaking-wave run-up point, because seawater infiltrates from the run-up point of breaking wave with semi-circular form as observed in the previous sections. To determine the vertical distribution of the lithium content, the sediment was sliced with 1 cm interval and mixed completely prior to extraction. The samples were transported in test tube and acidified with acid (HNO_3) of 0.01-N concentration and then vigorously shaken for 10 min. The extracts were centrifuged at 5000 rpm for 15 min and analyzed for lithium with

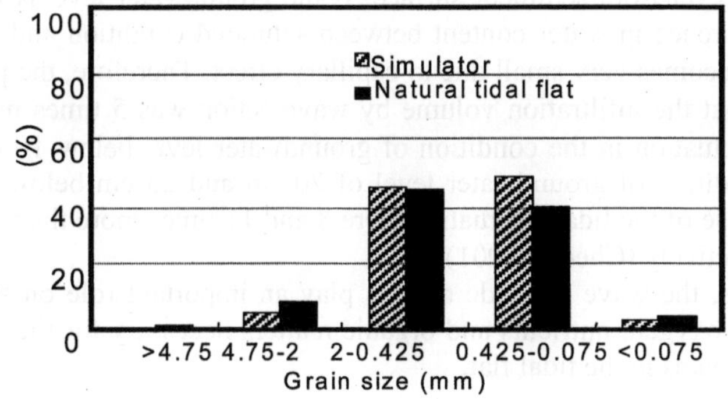

Fig. 2.8. Grain size distribution of tidal flat sediment

Inductively Coupled Argon Plasma Automatic Emission Spectrometer ICP (ICAP-575 II, Nippon Jarrell-Ash).

Total infiltration volume V_{wave} ($m^3\ m^{-2}\ day^{-1}$) of seawater by wave action was calculated by the following formula.

$$V_{wave} = A \cdot W \cdot P \cdot t \cdot N / (W \cdot L)$$

where A is rate of infiltration area of seawater ($m^2\ sec^{-1}$), W is width of flat bed (m), P is porosity, t is elapsed time (sec) to move the water level to length 1 m, N is number of tidal excursion a day (4 times), L is length of flat bed (m), and infiltration area of seawater was given the plotted area which determined by the detection limit depth of lithium tracer in each point. Thus, determined infiltration volume was 0.45 x 10^{-3} m^3 m^{-2} day^{-1}. This value is very small compared with that for sandy beach because of large amount of fine particles in sediment as shown in Fig. 2.8.

Seawater infiltration by tide action

Seawater infiltration by tidal fluctuation was defined by the difference in water content between saturated condition and unsaturated condition for 6 h. Sediment was sampled with a cylindrical acryl core (inner diameter 2.5 cm, length 60 cm) at intervals 0.5 h and dried at 105 ℃ for 24 h and then weighed. Infiltration volume V_{tide} (m^3 m^{-2} day^{-1}) of seawater by tidal fluctuation was calculated by using following formula.

$$V_{tide} = N \cdot \Delta W / (\rho \cdot A_c)$$

where N is number of tidal excursion a day (2 times, day^{-1}), ΔW is difference in water content between saturated condition and unsaturated condition (kg), ρ is density of seawater (kg), A_c is area of the core (m^2). The water content is depending on groundwater level from sediment surface. If the groundwater level is less than 10 cm, the difference in water content between saturated condition and unsaturated condition becomes very small due to capillary effect. Therefore, the present study showed that the infiltration volume by wave action was 5 times more than that of tidal fluctuation in the condition of groundwater level below 10 cm. However, in the condition of groundwater level of 20 cm and 25 cm below the sediment surface, those of the tidal fluctuation were 8 and 11 times more than that of wave action, respectively (Cheong 2001).

From these results, the wave and tide actions play an important role on supplement of dissolved oxygen, nutrients and organic matters necessary for the survival of benthic organisms in the tidal flat.

2.4.2 Observation of benthic organisms in tidal flat simulators

Experimental set-ups

Two simulators with the same configuration as shown in Fig.2.1 were used for this study. Natural tidal flat sediments and macrobenthos collected from a tidal flat in Hiroshima Bay, Japan, were placed in the simulators. In order to acclimate and stabilize benthic ecosystem, two tidal flats were operated for 70 days under the same tide and wave conditions. The tidal level was controlled as a semi-diurnal tidal cycle (12 h) with the rate of 0.009 cm sec^{-1}, *i.e.* the same vertical tidal fluctuations rate as that in Hiroshima Bay. Wave height and wave period were 30 mm and 0.8 sec, respectively. One tidal flat was used as a control (T&W simulator) with wave and tide. Another one (T simulator) was operated with wave and tide for 70 days and then the wave was stopped after day 70.

Sediment and seawater analyses

Sediment and seawater were prepared to have the same conditions as those in natural tidal flats. Physico-chemical characteristics of tidal flat sediments and seawater were evaluated by particle size, total organic carbon and inorganic nutrients. Sediment samples were washed, desalinated and after removal of organic matter with 30% hydrogen peroxide (H_2O_2), they were finally dried. The dried sediment samples were fractionated into pebble (> 4.75 mm), fine pebble (2.0 – 4.75 mm), coarse sand (0.425 – 2.0 mm), fine sand (0.075 – 0.425 mm) and silt + clay (< 0.075 mm) fractions using a sieve (Lee 1998).

After 49 days from the stop of wave, sediments were cored to determine vertical distributions of total organic carbon contents (TOC). Sediment samples were dried at 105℃ for 2 hours, and sieved (< 2 mm) to determine TOC by CN corder (Yanagimoto, MT-700). Seawater was sampled once a week from the reservoir to determine dissolved inorganic nutrients ((NO_2 + NO_3) – N, PO_4 – P). The seawater samples were filtered through a glass fiber filter (Whatman GF/C) and nutrients were analyzed according to the method described by Grasshoff et al. (1999) and Lee (1996). Sediment samples for chlorophyll-*a* determination were collected from the surface of the sediment. Chlorophyll-*a* concentration was analyzed according to Lorenzen (1967) after extraction with 90 % acetone overnight and centrifugation. To determine infiltration volume of seawater by tidal fluctuation, surface sediments were cored between flood and ebb tides. Infiltration volume of seawater by tidal fluctuation was estimated from the difference in water content in the sediments between flood and ebb tides according to the method already described in the previous section.

Population of macrobenthos and benthic algae

Sediment samples (L 20 × W 20 × D 10 cm) for macrobenthos population analyses were taken once a week. After sieving through 1mm mesh sieve, the samples were preserved by 10 % formalin with Rose Bengal for species identification. Macrobenthos were sorted and counted under a stereoscopic microscope in the laboratory (Tsutsumi 1990). Benthic algae was sampled from the surface of tidal flat (1 × 1 cm) and preserved by formalin. Algal cells of the surface of the sediments were identified and counted with the microscope.

Physico-chemical characteristics of the sediments

Silt and clay contents in T&W and T simulators were 2.87 %, 3.18 %, respectively. The silt content in natural tidal flats ranged from 2.71 to 8.68 % in Hiroshima Bay (Lee et al. 1998). Therefore, particulate size distribution of the tidal flat sediments in the simulators was almost the same as those in natural tidal flats. Inorganic nitrogen concentration was determined to identify the supply of nutrient with T&W and T simulators. $(NO_2 + NO_3) - N$ concentrations in seawater at T&W and T simulators ranged from 0.1 to 1.1 μM and from 0.1 to 1.5 μM, respectively (Fig. 2.9). The concentration of $(NO_2 + NO_3) - N$ in Hiroshima Bay ranged from 0.2 μM to 10 μM (Lee, 1996). The $NO_3 - N$ concentration was almost 59 – 99 % of DIN ($(NO_2 + NO_3) - N$, $NH_4 - N$) at low DIN concentration area in Hiroshima Bay (Lee 1996). Therefore, dissolved inorganic nitrogen concentrations in seawater were almost the same as those in Hiroshima Bay.

Fig. 2.10 shows inorganic phosphate ($PO_4 - P$) concentrations in seawater. The concentrations of inorganic phosphate at T&W and T simulators ranged from 0 to 0.26 μM and from 0 to 0.19 μM, respectively. These inorganic phosphate concentrations were similar to those in Hiroshima Bay (0.1 – 1.5 μM).

Fig. 2.11 shows vertical distributions of total organic carbon contents both in T and T&W simulators after 49 days from the stop of wave. Total organic carbon at 0 – 1 cm in T and T&W simulators were 1.27 % and 0.92 %, respectively. The higher TOC at the sediment surface in T simulator seems to be due to the growth of filamentous benthic algae on the surface of the sediments after the stop of wave.

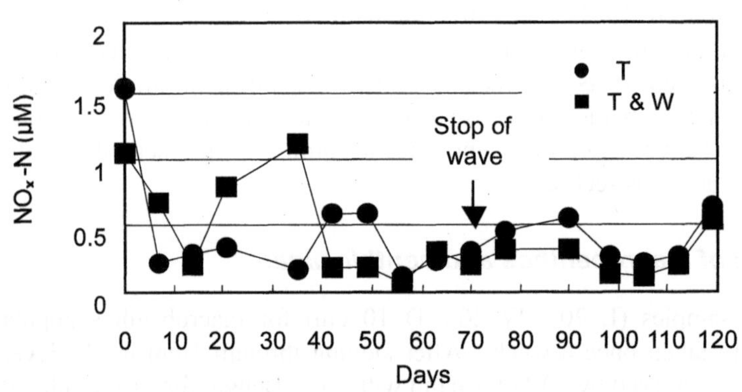

Fig. 2.9. Variations of (NO_2+NO_3)–N concentrations in seawater

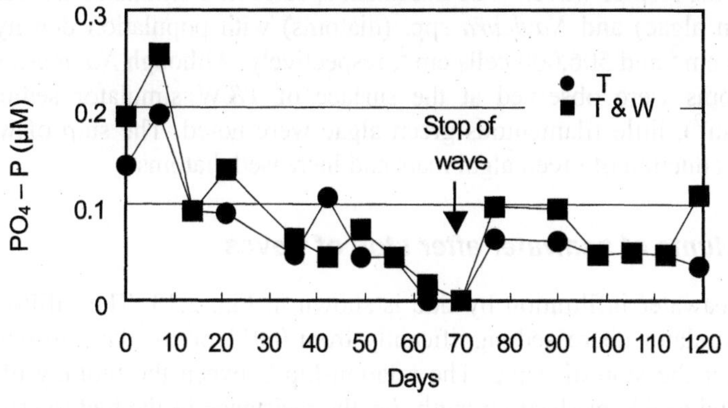

Fig. 2.10. Variations of PO_4–P concentrations in seawater

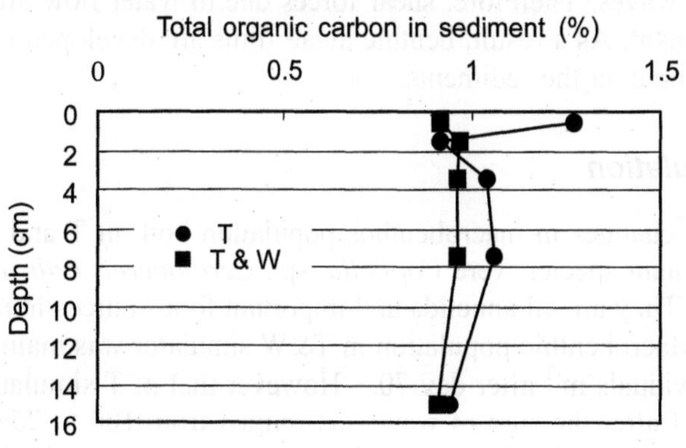

Fig. 2.11. Vertical profiles of total organic carbon (%) after 49 days from the stop of wave

Chlorophyll-a and Benthic algae

Fig. 2.12 shows chlorophyll-*a* concentration on the surface of tidal flat sediment. Chlorophyll-*a* concentration in the sediment ranged from 2.3 μg g^{-1} to 22.2 μg g^{-1} and the average for 70 days was 11.2 μg g^{-1} in T&W simulator. Those in T simulator ranged from 4.6 μg g^{-1} to 31.6 μg g^{-1} and the average was 10.1 μg g^{-1}. This suggests that chlorophyll-*a* concentration in two simulators showed similar trend until the stop of wave. Although chlorophyll-*a* concentration in T&W simulator slightly decreased after day 70, the concentration was remained almost the same range (3.1 μg g^{-1} – 9.5 μg g^{-1}). In T simulator, however, chlorophyll-*a* concentration increased more than 2 times after the stop of wave (16.7 μg g^{-1} – 31.6 μg g^{-1}). The higher concentration of chlorophyll-*a* at the surface of T simulator sediment seems to be due to the growth of benthic algae on the surface of the sediments after the stop of wave. The dominant species in T simulator were *Derbesia* sp. (green algae) and *Navicura* spp. (diatoms) with population density of 7,000 filaments cm^{-2} and 506,000 cells cm^{-2}, respectively. Although *Navicura* spp. and other diatoms were observed at the surface of T&W simulator sediment (30,000 cells cm^{-2}), little filamentous green algae were noted. The stop of wave enhanced the production of green algal mats and increased diatoms.

Infiltration volume of seawater after stop of waves

Volume of seawater infiltration by tide is shown in Fig. 2.13. The infiltration volume in T simulator decreased significantly from 0.011 $m^3\ m^{-2}\ day^{-1}$ to 0.0004 $m^3\ m^{-2}\ day^{-1}$ after the stop of wave. The relationship between the biology of the benthic fauna and sediment shear strength, *i.e.* the resistance of the bed to erosion (Hall, 1994) is a complex one. By stopping waves the flow fluctuation due to waves does not exist. On the other hand, flow movement by tidal motion itself is very slow than that of waves. Therefore, shear forces due to water flow are very small. It is almost stagnant. As a result, benthic algae films are developed and reduce seawater infiltration over the sediments.

Macrobenthos population

Fig. 2.14 shows the changes in macrobenthos population both in T and T&W simulators. The dominant species were *Capitella.* sp., *Ceratonereis erthraeensis* and *Mediomastus* sp. They are all annelida and important food sources in natural tidal flat ecosystems. Macrobenthic population in T&W simulator was maintained from 525 to 2400 individuals m^{-2} after day 70. However that of T simulator decreased less than a half after the stop of wave, *i.e.* ranged from 100 to 250 individuals m^{-2}. The remarkable reduction in population density was noted after the stop of wave in T simulator in spite of sufficient nutrients ($(NO_2 + NO_3) - N$, $PO_4 - P$) and chlorophyll-*a* in sediments and seawater, whereas T&W simulator kept high population density of macrobenthos. The reduction of macrobenthic population after the stop of wave is coincident with the growth of filamentous green algae and the decrease in the volume of seawater infiltration. The fronts of breaking waves always touch and lick the surface of sediments along with the tidal motions. Aerations due to wave breaking are also occurring at the front of breaking waves. Accordingly, shearing forces and aerations work on the surface of sediments. These phenomena over the sediment surface seem to be critical factors for the survival of macrobenthic community in tidal flats.

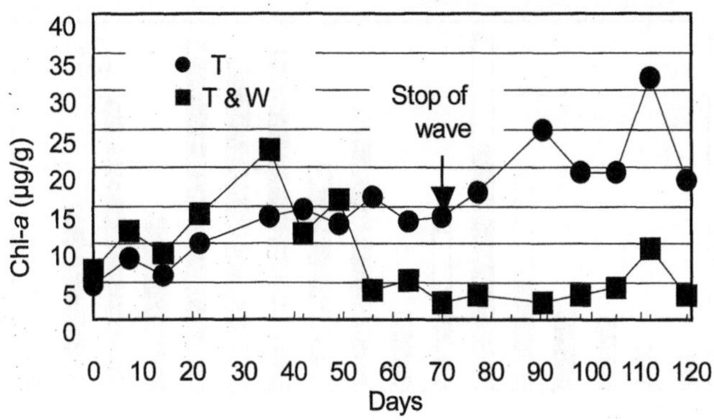

Fig. 2.12. Chlorophyll-*a* concentration in sediments at T and T&W simulators

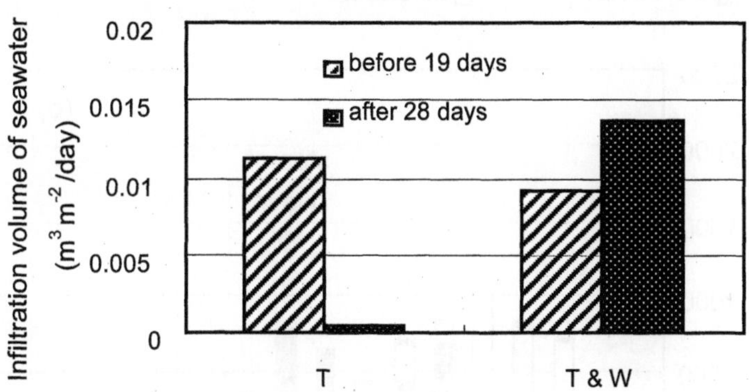

Fig. 2.13. Infiltration volume of seawater by tide at T and T&W simulators

This experimental result means that the ecosystem of tidal flat is changed with and without waves. Without waves a different ecosystem appears. It is understood that waves are indispensable for sound development of tidal flat where macrobenthic community live as the bases of food chain in seabed. This experimental finding is supported by an experimental result reported by Hosokawa and Kuwae who found that a periodic wave action over tidal flat activates growth of short-necked clams. In this study they used tidal flat facilities connected to Kurihama Bay at the entrance of Tokyo Bay. In the experiment, operations with and without waves were conducted: (Hosokawa and Kuwae 1997).

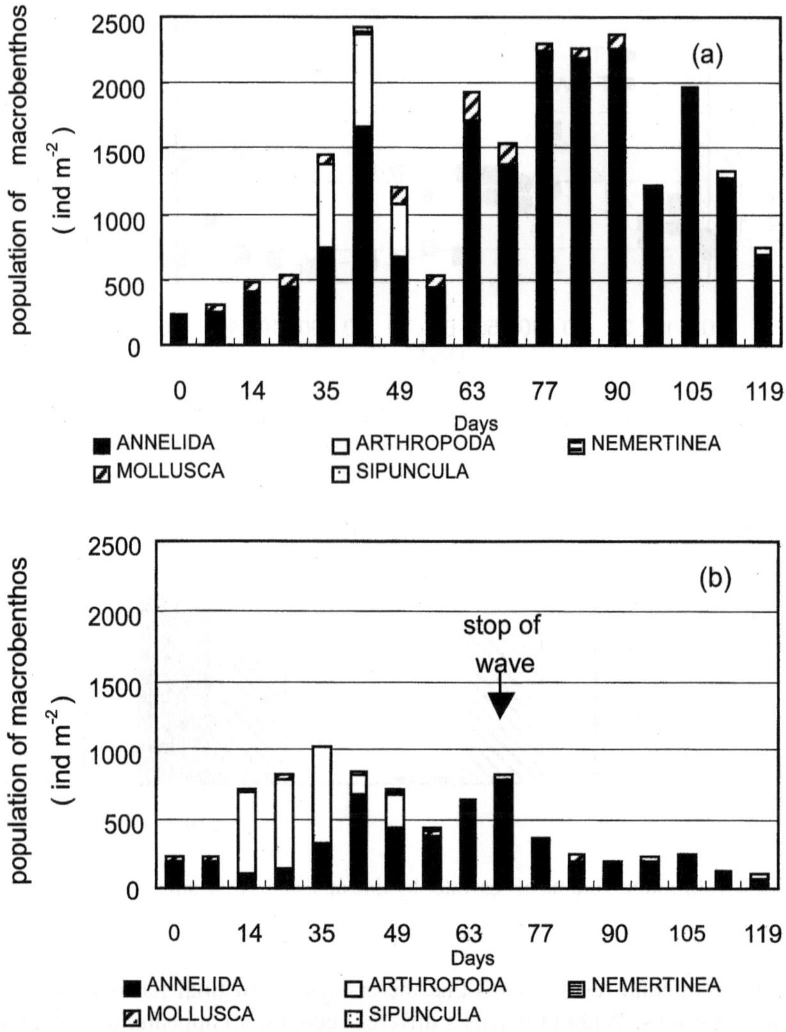

Fig. 2.14. Macrobenthic population in T&W (a) and T (b) simulators.

2.5 Concluding remarks

By using tidal flat experimental facilities, the internal flow of sandy beach and tidal flat under wave and tide actions was observed. At first, for flow visualization inside sandy beach, transparent glass beads were used instead of real soil from a tidal flat. Then, the real tidal flat sediment was used for the study on the role of waves and tides upon ecosystems of tidal flats. The following findings were obtained:

(1) Fresh seawater is pumped into seabed by breaking waves at the sloping seabed. The fresh seawater is transported radially from a point of near the run-up point of breaking waves.
(2) It became possible to quantify the water volume pumped by breaking waves by means of flow visualization technique using transparent glass beads in the experimental facility.
(3) For relatively steeper seabed, silt particles are sucked out from seabed at the point of wave breaking. This implies the decrease in silt content in the tidal flat of steeper slope. This observation coincides with results of field measurement shown in Chap.1.
(4) Waves play an important role on ecosystems of tidal flat. The density of dominant macrobenthic species, *Capitella* sp., *Ceratonereis erthraeensis* and *Mediomastus* sp., decreased significantly after the stop of wave.
(6) TOC contents and chlorophyll-*a* concentrations at the sediment surface increased after the stop of wave by the growth of filamentous benthic algae on the surface.
(7) The infiltration volume of seawater by tidal fluctuations also decreased significantly after the stop of wave. The remarkable reduction of macrobenthic population after the stop of wave was coincident with the growth of filamentous green algae and the decrease in the volume of seawater infiltration. Therefore, fresh seawater transport and the shearing stress to sediments surface by waves seems to be a critical factor for the survival of macrobenthic community in tidal flats.

References

Cheong CJ (2001) Penetration Behavior of Spilled Oil and its Effects on Coastal Zone Ecosystem. Ph. D. thesis, Department of Environmental Science, Faculty of Engineering, Hiroshima University

Cheong CJ, Lee JG, Nishijima W, Baba E, Okada M (2000) Seawater Infiltration into tidal flat or sandy beach sediment by wave action in enclosed bay. J Japan Society on Water Environment 23: 619-623

Doi Y, Higo Y, Baba E, Goto K (2000) Study on wave induced ground water flow in a surf zone (in Japanese). J Society of Naval Architects of Japan 188:33-39

Grasshoff K, Kremling K, Ehrhardt M (eds) (1999) Methods of Seawater Analysis. 3rd edn Weinheim, New York, Chichester, Brisbane, Singapore, Toronto

Hall SJ (1994) Physical disturbance and marine benthic communities: Life in unconsolidated sediments. Oceanography and Marine Biology 32:179-239

Hosokawa Y, Kuwae T (1997) Mesocosm Experiment by Tidal-Flat Expeiment Facility (in Japanese) J Japan Society of Civil Engineers 82:12-14

Lee YS (1996) The mechanism of phytoplankton growth in Hiroshima Bay. Ph.D. thesis, Department of Environmental Science, Faculty of Engineering, Hiroshima University

Lee JG, Nishijima W, Mukai T, Takimoto K, Seiki T, Hiraoka K and Okada M (1998) Factors to determine the functions and structures in natural and constructed tidal flats. J Water Research 32(9): 2601-2606

Lee JG, Cheong CJ, Nishijima W, Okada M, Baba E (1999) Effect of seawater flow on the soil structure in tidal flat (in Japanese). J Japan Society of Water Environment 22:677-682

Longuet-Higgins MS (1983) Wave set-up, percolation and undertow in the surf zone. Proc. Royal Society London A390: 283-291

Lorenzen CJ (1967) Determination of Chlorophyll and pheopigments: spectrophotometric equations. Limnol Oceanogr 12: 343-346

McLachlan A (1982) A model for the estimation of water filtration and nutrient regeneration by exposed sandy beaches. Marine Environmental Research 6:37-47

Riedel RJ (1971) How much seawater passes through sandy beach? Internationale Revue der Gesamten Hydrobiologie 56:923-946

Sorensen M (1997) Two-dimensional wave equations and wave characteristics. In: Basic Coastal Engineering 2nd eds. Chapman and Hall, pp 9-52

Tsutsumi H (1990) Population persistence of *Capitella* sp. (Polychaeta; Capitellidae) on a mud flat subject to environmental disturbance by organic enrichment. J Mar Ecol Prog Ser 63: 147-156

Wrenn BA (1997) Nutrient transport during bioremediation of contaminated beaches: evaluation with lithium as a conservative tracer. J Water research 31(3):515-524

3 Unified model for wave breaking action and internal flow in a tidal flat

3.1 Introduction

What kind of correlation exists between waves on the beach and tidal flat ecosystems? This chapter clarifies the above problem using mathematical science. Concretely, we simulate phenomena involved in the problem by building a mathematical model.

Generally speaking, the meaning of numerical simulation comprises: (1) Clarification of mechanisms for governing systems, (2) Realization of artificial circumstances by altering physical parameters, (3) Prediction of future states of governing systems, (4) Control of governing systems for more desirable outcomes, and so on.

Experimental observations with respect to internal flow of a tidal flat under wave and tide actions are summarized as follows:

I. Fresh seawater infiltration
   a) Seawater infiltration into the seabed in a semicircular pattern around one point near the run-up point due to breaking waves.
   b) Quantification of seawater infiltration due to wave action by integrating the semicircular area, making it possible to evaluate seawater infiltration under the effect of spilled oil.

II. Silt movement by wave action

   For relatively steeper seabed slopes, silt particles are sucked up from the seabed at the point of wave breaking, implying a decrease in silt content in the tidal flat with a steeper slope. This observation coincides with field measurement results.

We can conceptualize the following physical characteristics from I and II above.

i. Two phase flow consisting of air and seawater in various regions such as air, water, wet sand, and dry sand limited by the sloping beach. The interface between air and seawater phases is free.
ii. Internal flows of air and water in the sandy beach receive resistance force proportional to velocity of flow from sand, whose coefficient is assumed to be an inverse of porosity.
iii. When seawater crosses the surface of a tidal flat covered with accumulated materials, it receives resistance force due to friction exerted at the surface.

We try to realize observations I and II by computer through the following process:

(1) Constructing mathematical models by aid of applicable sciences.
  - Conservation of mass for a total flow system.
  - Conservation of momentum for a total flow system.
  - Surface blocking effect to seawater due to accumulated materials on the beach.
(2) Construction of a numerical model for the mathematical model by unifying flow equations of all phases due to the distribution theoretic approach.
(3) Discretization of the numerical model based on finite difference approximation.
(4) Development of solution algorithms to solve the discretized numerical model.
  - Discretization effect for an incompressibility condition in two-phase flow with large density discrepancy.
  - An anti-smearing device for numerical free surface.
(5) Visualization of numerical results.
  - Various visualization methods to understand mechanisms of the concerned system.
  - Animated graphics to understand transition of unsteady phenomena.

Through the above process, we want to observe the following:

(A) Correlation between wave motion on a sloping beach and internal flow of seawater in sand, especially specification of penetration and suction points of seawater on the beach and clarification of internal flow pattern and infiltration area which contribute to biological activity.
(B) Circulation flow in sand on the slope for a tidal flat or sandy beach, which has a close relation with the outflow of silt particles from the beach: the circulation flow has a strong correlation to the amount of silt and the infiltration area of underground seawater. The search for an appropriate value of the slope to balance lessened outflow of silt and greater enlargement of flow area will suggest guidelines to build a man-made tidal flat.

3.2 Mathematical model

3.2.1 Notations and geometry

x_i : Cartesian coordinates $(i = 1, 2, 3)$
u_i : Velocity vector $(i = 1, 2, 3)$
u_n : Normal component of velocity on boundary
u_T : Tangential component of velocity on boundary
p : Pressure
ρ : Density
t : Time

D_{ij}	: Rate of strain tensor
Ω_α	: Domain for α-phase, where $\alpha = \{air,\ water\}$
$\Omega = \bigcup\limits_\alpha \overline{\Omega_\alpha}$	: Total domain
Ω_L	: Domain occupied with sand
Ω_U	: $\Omega\backslash\overline{\Omega_L}$
χ_α	: Characteristic function representing domain Ω_α, i.e., $\chi_\alpha = 1$ in $\overline{\Omega_\alpha}$ and $\chi_\alpha = 0$ in $\Omega\backslash\overline{\Omega_\alpha}$
χ_L	: Characteristic function representing domain Ω_L,
χ_U	: Characteristic function representing domain Ω_U,
χ_α^L	: $\chi_L \cdot \chi_\alpha$,
χ_α^U	: $\chi_U \cdot \chi_\alpha$,
ν^α	: Kinematic viscosity of α-phase
μ^α	: Viscosity of α-phase
σ_{ij}^α	: Stress tensor of α-phase
ρ^α	: Density of α-phase
c	: Coefficient of resistance force received from sand
g_n	: Friction coefficient of accumulated material
K_i	: ith component of external force
$\|A\|$	: Euclid norm of A
Γ_B	: Boundary between Ω_{water} and Ω_{air}
Γ_I	: Inflow boundary
Γ_S	: Slip boundary
Γ_N	: No-slip boundary
Γ_A	: Outflow boundary

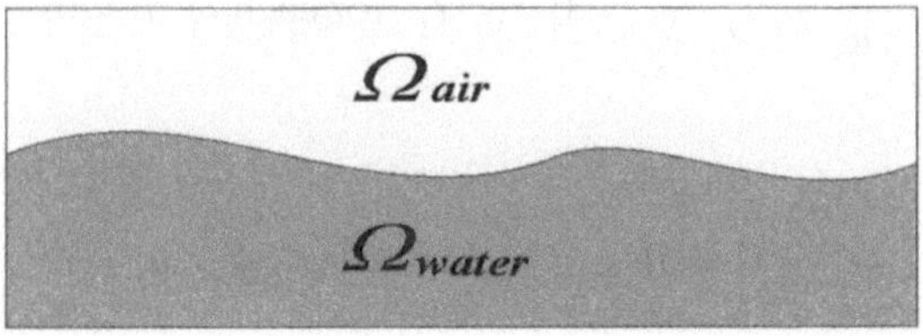

Fig. 3.1. Two-phase flow

Figure 3.1 shows Ω_α where $\alpha = \{air,\ water\}$ and Fig. 3.2 shows Ω_L and Ω_U. In this problem, the domain Ω is consists of four parts.

1 The first part ($\Omega_{air} \cap \Omega_U$) is occupied with air.
2 The second part ($\Omega_{water} \cap \Omega_U$) is filled with water.
3 The third part ($\Omega_{water} \cap \Omega_L$) is occupied with water penetrated into sand from the sloping beach.
4 The fourth part ($\Omega_{air} \cap \Omega_L$) is the dry part of sand filled with air.

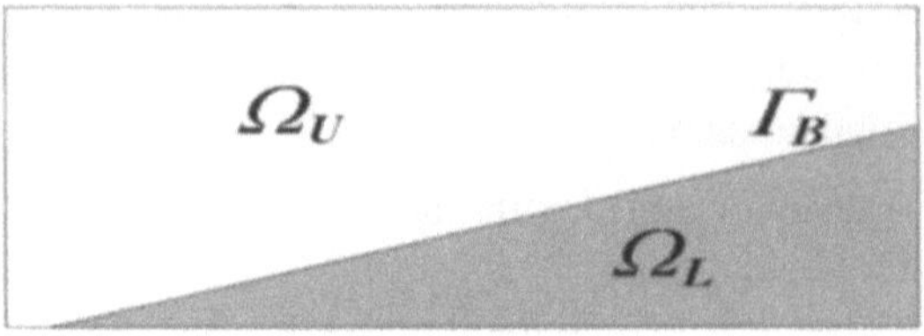

Fig. 3.2. Sloping beach

3.2.2 Conservation of mass for total flow system

Total mass of two-phase flow system is conserved;

$$\frac{\partial \rho}{\partial t} + \frac{\partial}{\partial x_j}(\rho u_j) = 0, \quad \text{in } \Omega. \tag{3.1}$$

Here, $\rho = \sum_l \rho_l \chi_l, \quad u_i = \sum_l u_i^l \chi_l$, and l represents $\{water, air\}$.

Fluids for each phase are assumed to be incompressible, i.e., ρ_l = const. Then we have

$$\sum_l \rho_l \left\{ \frac{\partial \chi_l}{\partial t} + u_j^l \frac{\partial \chi_l}{\partial x_j} \right\} + \sum_l \rho_l \chi_l \frac{\partial u_j^l}{\partial x_j} = 0 \ \text{ in } \Omega, \ t > 0. \tag{3.2}$$

From (3.2) and kinematic conditions for interface between different phases,

$$\frac{\partial \chi_l}{\partial t} + u_j^l \frac{\partial \chi_l}{\partial x_j} = 0 \ \text{ in } \Omega, \quad \text{for each } l, \quad t > 0. \tag{3.3}$$

It follows that,

$$\frac{\partial u_j^l}{\partial x_j} = 0 \ \text{ in } \Omega_l, \ \text{for each } l, \quad t > 0, \tag{3.4}$$

which shows the incompressibility condition for fluids in each phase.

3.2.3 Conservation of momentum for total flow system

Equations of motion for seawater and air Motion equations for seawater and air on the beach are described as Navier-Stokes equations:

$$\frac{\partial u_i}{\partial t} + u_j \frac{\partial u_i}{\partial x_j} = -\frac{1}{\rho_l} \frac{\partial p}{\partial x_i} + \nu_l \Delta u_i + K_i. \tag{3.5}$$

Motion equations for seawater and air in sand are represented by ones including the resistance force $-cu_i$ from the sand. Here, c is proportional to the inverse of sand porosity.

The unified equations of motion for two-phase flow Conservation of momentum for the total system of two-phase flow is represented using motion equations for each flow.

$$\rho\left(\frac{\partial u_i}{\partial t}+u_j\frac{\partial u_i}{\partial x_j}\right)=$$

$$-\frac{\partial p}{\partial x_i}-c\chi_L u_i+\rho K_i+\frac{\partial}{\partial x_j}\left(\mu D_{ij}\right)\qquad \text{in } \Omega,\ \ t>0, \tag{3.6}$$

$$\frac{\partial u_j}{\partial x_j}=0\qquad \text{in } \Omega,\ \ t>0, \tag{3.7}$$

where $\mu=\mu_{air}\chi_{air}+\mu_{water}\chi_{water}$.

3.2.4 Surface blocking effect to seawater due to accumulated materials on the beach

Here we discuss the effect brought about by accumulated materials on the surface of a tidal flat to the internal flow. Seawater flowing into both sides of the surface receives a resistance force from accumulated materials. We propose a mathematical model for this phenomenon due to Signorini-type boundary conditions: (Glowinski et al. 1981)

Let the jump between normal components of the stress for water of both sides at the surface be $[\sigma_n]$. If the absolute value of $[\sigma_n]$ remains within the threshold based on resistance, water does not pass through the surface. If this jump increases and reaches the threshold, water starts to penetrate into both sides. Direction of flow depends upon the sign of the jump. These phenomena are described by means of the unilateral boundary condition for the jump $[\sigma_n]$ as follows:

$$\begin{cases} |[\sigma_n]|\le g_n, \\ (|[\sigma_n]|-g_n)\cdot u_n=0, \\ [\sigma_n]\cdot u_n\le 0, \end{cases}\qquad \text{on } \Gamma_B, \tag{3.8}$$

where g_n represents the threshold of penetration. Large magnitude of the threshold corresponds to impenetrable material for seawater. On the other hand, a small one implies permeable material. The second relation in (3.8) means the following:

$$\begin{cases} if\ \ |[\sigma_n]|<g_n,\ \ then\ u_n=0, \\ if\ \ |[\sigma_n]|=g_n,\ \ then\ \begin{cases} u_n=0, \\ u_n\neq 0\ \ \left([\sigma_n]=-g_n\dfrac{u_n}{|u_n|}\right). \end{cases} \end{cases} \tag{3.9}$$

Let us note that the jump $[\sigma_n]$ satisfying the unilateral condition is expressed by

$$[\sigma_n]=-g_n\cdot\partial\left(|u_n|\right), \tag{3.10}$$

where $\partial(\cdot)$ means a subdifferential. Equation (3.10) defines a stress boundary condition with a jump prescribed at the beach surface. In this study, this condition is regularized in order to avoid numerical singularity as

$$[\sigma_n] = -g_n \cdot \frac{u_n}{\sqrt{u_n^2 + \varepsilon_g^2}}, \tag{3.11}$$

where ε_g is a small positive parameter: (Kawarada and Suito 1997)

3.3 Unified model for two-phase flow with surface blocking effect to seawater on the beach

The incompressibility condition for two-phase flow system is

$$\frac{\partial u_j}{\partial x_j} = 0, \quad \text{in } \Omega, \quad t > 0. \tag{3.12}$$

Conservation of momentum for the total flow system is established as a unified equation:

$$\rho\left(\frac{\partial u_i}{\partial t} + u_j \frac{\partial u_i}{\partial x_j}\right) = -\frac{\partial p}{\partial x_i} - c\chi_L u_i + \rho K_i + \frac{\partial}{\partial x_j}(\mu D_{ij})$$
$$+\chi_{water} g_n \frac{u_n}{\sqrt{u_n^2 + \varepsilon_g^2}} \frac{\partial \chi_L}{\partial x_i} \quad \text{in } \Omega, \quad t > 0. \tag{3.13}$$

The last term in the right-hand side of (3.13) represents penetration resistance occurring at the surface of the beach. Introduction of this term is based on the distribution theoretic approach to multi-phase flow. (See Chap. 9.)

The phase equation is

$$\chi_{water} + \chi_{air} = 1 \ \text{ in } \Omega, \ t > 0. \tag{3.14}$$

The motion equation of free surface for air is

$$\frac{\partial \chi_{air}}{\partial t} + u_j \frac{\partial \chi_{air}}{\partial x_j} = 0 \ \text{ in } \Omega, \ t > 0. \tag{3.15}$$

Finally, outer boundary conditions are prescribed as

$$\begin{cases} u_n = u_0 \sin \omega t, \\ u_T = 0, \end{cases} \text{on } \Gamma_I(t), \tag{3.16}$$

$$\begin{cases} u_n = 0, \\ \dfrac{\partial u_T}{\partial n} = 0, \end{cases} \text{on } \Gamma_S, \tag{3.17}$$

$$\begin{cases} u_n = 0, \\ u_T = 0, \end{cases} \text{on } \Gamma_N, \tag{3.18}$$

$$\begin{cases} \sigma_n = 0, \\ u_T = 0, \end{cases} \text{on } \Gamma_A, \tag{3.19}$$

where ω and u_0 are the angular velocity and amplitude of incoming waves, respectively. We intend to clarify physical properties stated in I and II in Section 3.1 by seeking, numerically, the unknowns $\{u_i, p, \chi_\alpha\}$ for the unified model problem.

3.4 Numerical results

3.4.1 Correlation between wave breaking action on a sloping beach and internal flow of a sandy beach

Figures 3.3 to 3.5 show time evolution of free surfaces between water and air and the velocity field in sand after wave breaking. Locations of points Z_B (wave breaking point), $Z_R(t)$ (wave front), Z_{MR} (maximum run-up point), $Z_P(t)$ (strong penetration point) and Z_S (suction point) are shown in these figures, respectively. They imply, symbolically, characteristics of correlation between waves and internal flows.

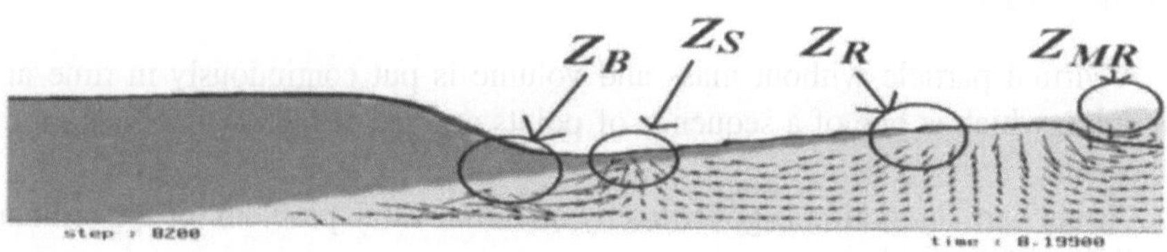

Fig. 3.3. Wave breaking

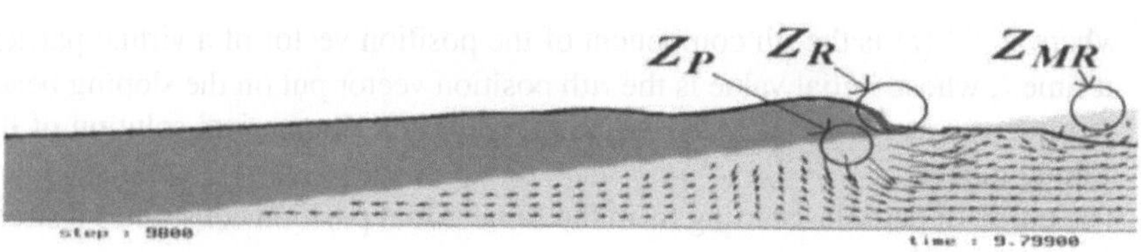

Fig. 3.4. Wave run-up after breaking

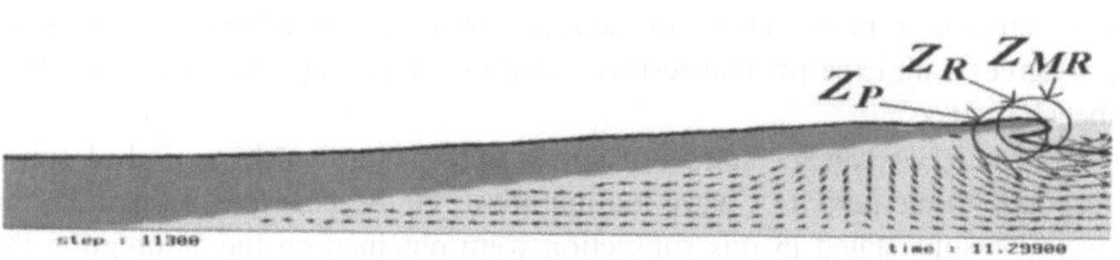

Fig. 3.5. Wave-front reaches the maximum run-up point

Figure 3.3 depicts the wave breaking, after which the wave runs up along the beach (Fig. 3.4). Figure 3.5 shows that $Z_R(t)$ reaches Z_{MR}. It can be observed that the water penetrates into the sand at a slightly lower region than $Z_R(t)$ at any time after the wave breaking until $Z_R(t)$ reaches Z_{MR} and water flows out from Z_S.

Numerically computed stress shows that the pumping effect induced by stress due to the wave breaking allows water to penetrate into the sand region.

3.4.2 Flow pattern and infiltration area of seawater in sandy beach

In order to see internal flow behaviour in a tidal flat, we visualize the flow field by drawing streak lines which are defined as curves linking the fluid particles continuously started from the fixed point x^0 on the beach. Streak lines are drawn in the following order:

1 A virtual particle without mass and volume is put continuously in time at a point, which is one of a sequence of points arrayed at regular intervals on the beach.
2 Virtual particle motion is governed by an initial value problem for the ordinary differential equation:

$$\begin{cases} \dfrac{dX_i^{m,n}(t)}{dt} = u(X_i^{m,n}(t), t), \quad (t > m\Delta t), \\ \qquad \text{for } n = 1, 2, \cdots, N \text{ and } m = 0, 1, \cdots, M, \\ X_i^{m,n}(m\Delta t) = x_i^{n0}, \quad (x_i^{n0} \in \Gamma_B), \end{cases} \tag{3.20}$$

where $X_i^{m,n}(t)$ is the ith component of the position vector of a virtual particle at time t, whose initial value is the nth position vector put on the sloping beach and is given at $t = m\Delta t$. Also, $u_i(\mathbf{x}, t)$ represents a numerical solution of the unified model.
3 The union of particles starting from the same initial position constitutes a streak line.
4 Tips of streak lines indicate approximate depth of seawater infiltration into sand.

Figure 3.6 shows streak lines, which are obtained by solving (3.20) numerically.

An envelope curve of streak line tips forms a semi-circle centered at a slightly lower level than the maximum run-up point Z_{MR}; from this, an aspect of the transport of nutrients from the surface of the sandy beach can be estimated. This observation agrees with experimental results stated in Chap. 2 by introducing dye from the beach surface.

In Fig. 3.6, we can see spirals of streak lines. Note that the period of spirals corresponds to the period of wave motion above the beach. Here, let us note that numerical results stated in this subsection were obtained on the assumption that $g_n = 0$, i.e., without accumulated materials on the beach.

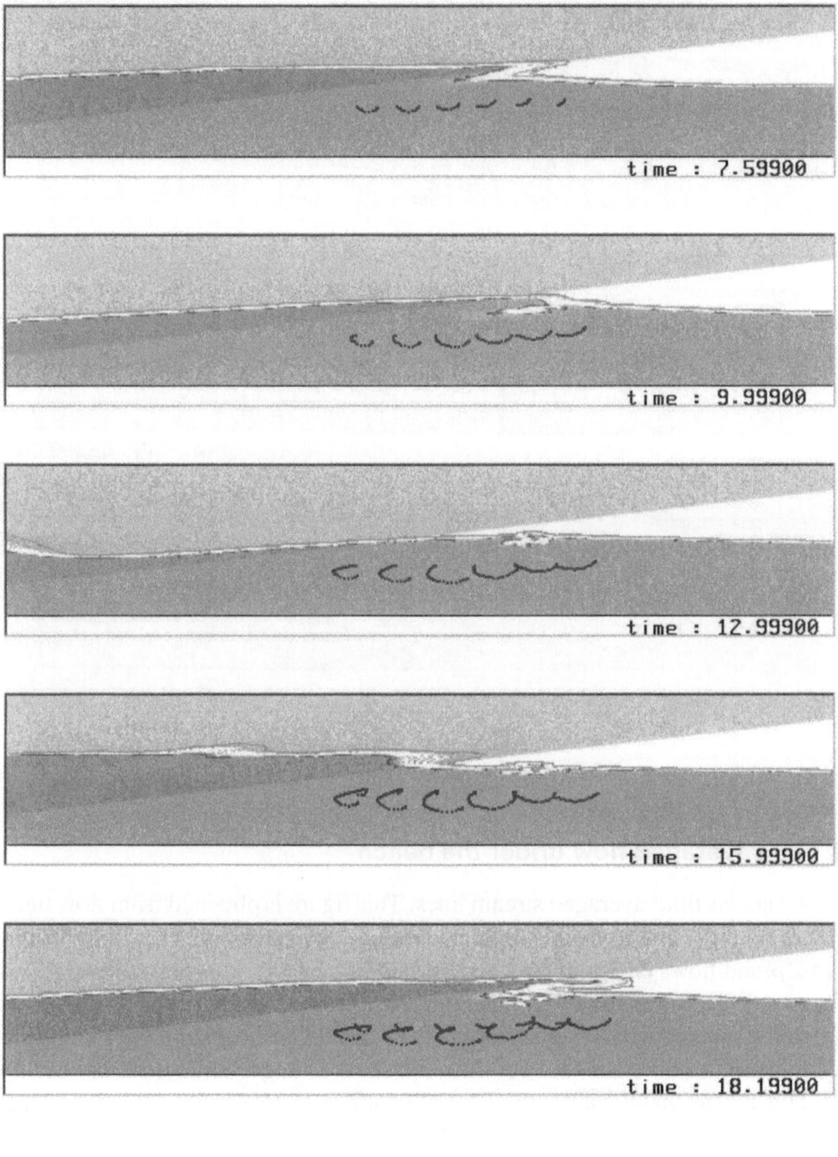

Fig. 3.6. Streak lines

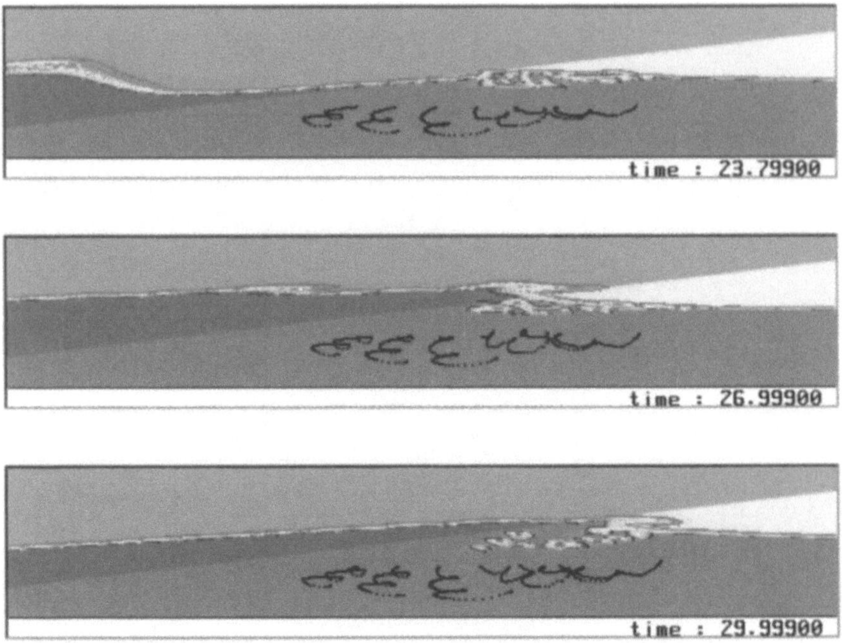

Fig. 3.6. Streak lines (continued)

3.4.3 Time averaged flow under the beach

Figure 3.7 shows time-averaged stream lines. This figure is obtained from flow fields which are averaged for a one wave period. We can see that seawater penetrates into sand at Z_P and flows out from the neighborhood of Z_S.

Figures 3.8 and 3.9 show the time-averaged normal component of velocity on the beach for cases in which beach slopes are 6° and 3°, respectively. They show that the total amount of flow into and out from the beach is larger for larger slopes of the beach. This means that the amount of the silt outflow is greater for the larger slopes of the beach. This results in decrease of the bacterial population, which agrees well with experimental results stated in Chap. 2.

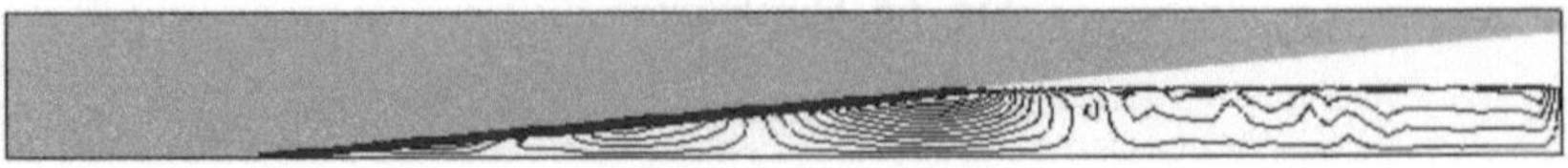

Fig. 3.7. Time-averaged stream lines under the beach

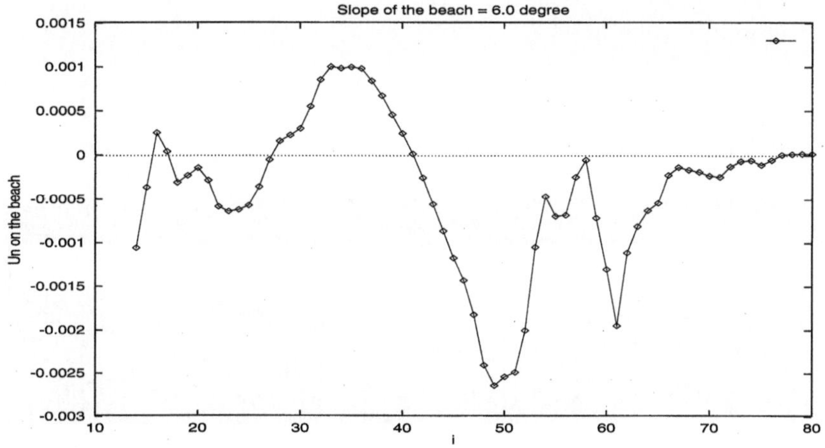

Fig. 3.8. Time-averaged normal component of velocity on the beach (slope = 6°)

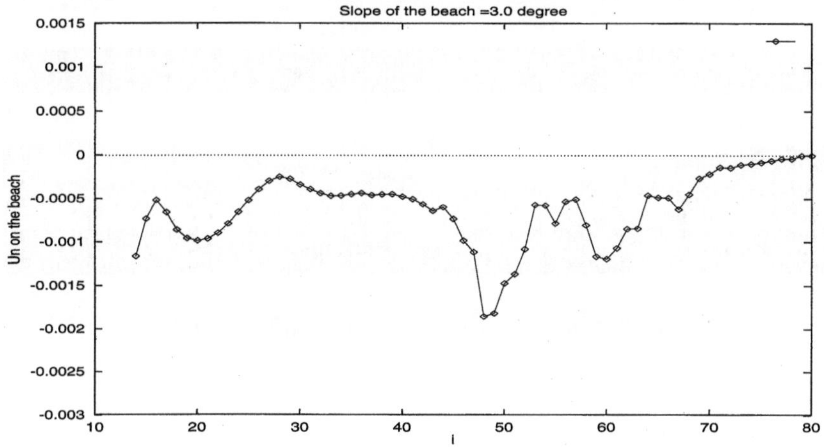

Fig. 3.9. Time-averaged normal component of velocity on the beach (slope = 3°)

3.4.4 Surface blocking effect to seawater

It is possible to let the coefficient g_n of penetration resistance at the beach relate to the quality of accumulated materials. Large (small) values of g_n correspond to the material with large (small) resistance for penetration of water meaning that reduction (enlargement) of the infiltration area; that same area is the habitat of ecosystem. For example, if beach surface is covered by oil (sediment), then large (small) g_n should be chosen. Here, we again apply streak line visualization to the case of an impenetrable surface of the beach. We show the infiltration area for several pairs $\{g_n, c\}$ in Figs. 3.10 and 3.11.

Fig. 3.10. Seawater penetration into sand ($g_n = 0.0$)

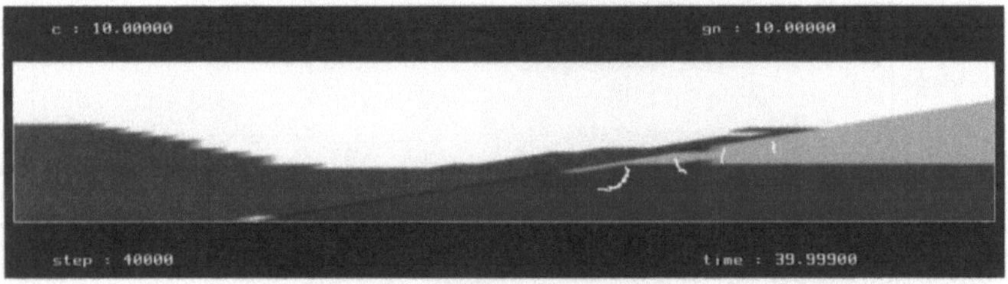

Fig. 3.11. Seawater penetration into sand ($g_n = 10.0$)

References

Fujita H, Kawarada H (1998) Variational inequalities for the Stokes equation with the boundary conditions of friction type. GAKUTO International Series Mathematical Sciences and Applications 11:15–33

Fujita H, Kawahara H, Kawarada H (1995) Distribution theoretic approach to fictitious domain method for Neumann problems. East-West J Numer Math 3:111–126

Glowinski R, Lions JL, Tremolieres R (1981) Numerical Analysis of Variational Inequalities, North-Holland Publishing Company.

Kawarada H, Fujita H, Suito H (1998) Wave motion breaking upon the shore. GAKUTO International Series Mathematical Sciences and Applications 11:145–159

Kawarada H, Suito H (1997) Numerical method for a free surface flow on the basis of the fictitious domain method. East-West J Numer Math 5(1):57–66

4. Oil pollution: human damages on hydraulic regime in sandy beach ecosystems

4.1 Introduction

Various anthropogenic pollutants ultimately threaten coastal areas. One of the most significant anthropogenic pollution is oil pollution. It is estimated that 1.7-8.8 × 10^6 tons of petroleum hydrocarbon are released into marine environment annually (Natural Research Council, 1985). Oil enters marine environment in many ways like natural seeps, refinery emission, ship cleaning operations and accidental spills. In particular, tanker accidents that may result in the release of relatively large amount of oils near sensitive coastal environments are of great concern. For example, the *Amoco Cadiz* discharged 0.2 × 10^6 tons of Kuwait crude oil into the waters along the Brittany coast in March 1978; the *Exxon Valdez* released 0.04 × 10^6 tons of Alaskan North Slope crude oil into Prince William Sound in March 1989. A part of the spilled oil into sea has drifted to the coastal area. About 30% of the *Amoco Cadiz* spilled oil contaminated 320 km of coastline. In case of the *Exxon Valdez,* about 50% of the spilled oil stranded on 2,000 km of shorelines along the Gulf of Alaska (Swannell *et al.*, 1996; Wolfe *et al.*, 1994; Vandermeulen *et al.*, 1979).

When spilled oil stranded and penetrated into coastal sediments, the oil may deteriorate benthic ecosystem. The primary mechanisms by which petroleum hydrocarbons are deleterious to benthic organisms include oxygen stress through turning the sediment anaerobic by organic pollution (Jhonston, 1970) and direct toxic or carcinogenic effects on organisms (Connell & Miller, 1984) like the reduction of grazing pressure of grazers (Carman *et al.*, 2000). The ecological damages may be caused not only by the toxicity of oil constituents but also by the changes in physico-chemical conditions like infiltration of seawater.

Macrobenthic organisms living in shoreline area obtain their food such as plankton, bacteria and detritus mainly from seawater. Wave action and/or tidal fluctuation in the shoreline ecosystem is known to play an important role in seawater infiltration processes during which dissolved and particulate organic matter is transported and then mineralized by the interstitial fauna as it returns to the sea by gravity drainage (McLachlan, 1982; Cheong *et al.*, 2000). It is likely that stranded oil blocks interstitial spaces of sediments, reduces water infiltration, and results in the decrease in oxygen, nutrients and other food supply to benthic communities. The depth of oil penetration into the sediments is one of the most significant factors to change biodegradation processes and/or physical dispersion (Owens, 1978).

In order to minimize the deterioration of shoreline ecosystem, it is necessary to remove the oil as quickly as possible. Cleaning techniques of the penetrated oil include manual pick-up with absorbent pads, mechanical treatment using equipment such as grader and scraper, chemical treatment by use of dispersants, digging trenches and bioremediation (Cormack, 1999). Prior to adopt any cleaning technique, it is necessary to know the penetration behavior of spilled oil into the sediments.

This chapter deals with the effects of wave and tidal actions on the penetration of spilled oil into sandy beaches and tidal flats and to evaluate the influence of the penetrated oil on seawater infiltration and benthic ecosystems. We made visualization of infiltration process of seawater and fuel oil C into sediments by using the same tidal flat simulators as shown in Chapter 2.2.1 with wave and tide controls and evaluated the effect of the penetrated oil on seawater infiltration into the sediments and on the benthic ecosystems.

Table 4.1. Physico-chemical properties of the tested crude oil and fuel oil C

	Crude oil	Fuel oil C
Viscosity at 15℃ ($mm^2\ s^{-1}$)	28	3,750
Density at 15℃ ($g\ cm^{-3}$)	0.87	0.95
Pour point (℃)	-15	-10
Sulfur content (% wt)	2.00	2.15
Water content (vol)	0.1	0.1

4.2 Infiltration of stranded oils into sandy beach sediments by waves and tides

4.2.1 Waves and Tides

Crude oil and fuel oil C shown in Table 1 were assumed to strand on a sandy beach with low energy. Breaking wave height and wave period were 30 mm and 0.8 sec, respectively. Breaking wave generated by the wave maker is classified as collapsing type (Sorensen, 1997). Tidal velocity was controlled at 0.009 cm · sec^{-1} assuming the tidal range of 2 m (=mean tidal range in Hiroshima Bay). The slope of the tidal flat was about 10/100.

Wave. Infiltration of fuel oil C by wave action is shown in Fig.4.1 (top). As shown in Chapter 2 (fig. 2.2), periodic wave forced infiltration of seawater into the sediments indicating waves were a driving force to transport seawater into the sediments. However, different from seawater, fuel oil C and did not strand on the beach nor penetrate into the sediments in spite of the wave action. The differences in the infiltration behaviors between seawater and oil would be due to the difference in specific gravity and viscosity.

Tide. Infiltration of fuel oil C by tidal fluctuation is shown in Fig.4.1 (middle). Both seawater and fuel oil C infiltrated into the sediments and flow down to the low seawater level during falling tides. The fuel oil C, however, penetrated with much slower velocity than seawater. The infiltration depths of seawater and fuel oil C in 1,500 seconds of a falling tide were approximately 10 cm and 2.5 cm, respectively. It is clear that fuel oil C penetrates into the sediments by tidal fluctuation and not by wave action.

Wave+Tide. Infiltration of fuel oil C by wave and tidal actions is shown in Fig.4.1 (bottom). Fuel oil C showed almost the same penetration as the tidal fluctuation without wave. The infiltration depths of seawater and fuel oil C after 1500 sec were about 18 cm and 2.5 cm, respectively.

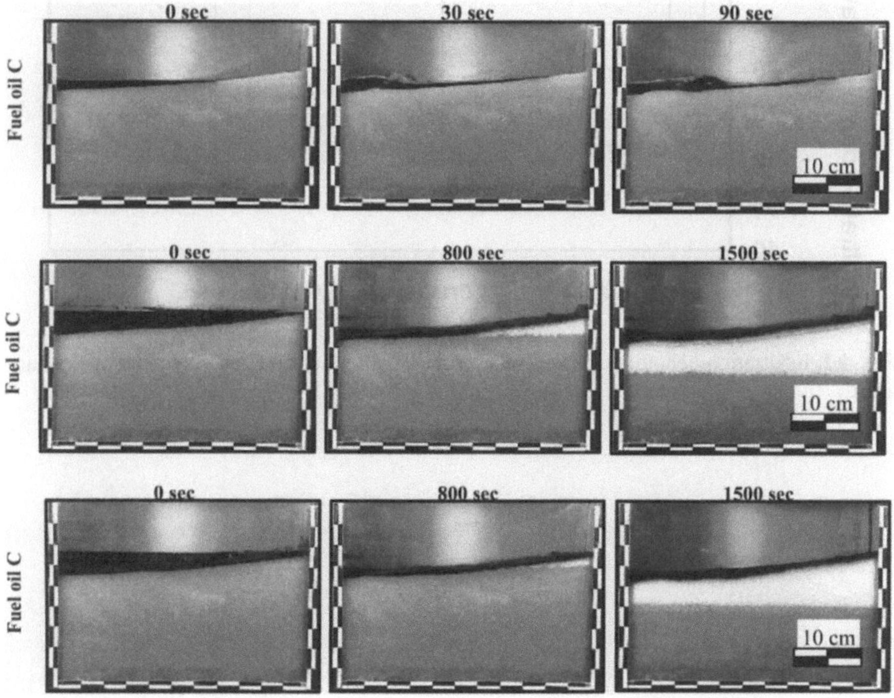

Fig.4.1. Infiltration of fuel oil C into simulated sandy beach (glass beads) by waves and/or tides

Fig. 4.2 shows penetration depths of seawater, crude oil and fuel oil C by wave and/or tidal actions. Seawater infiltrated into the sediments to 21 cm from the surface by wave actions for 5 min, whereas crude oil and fuel oil C did not penetrate into the sediments. Seawater also infiltrated into the sediments by tidal fluctuation down to 30 cm in 6 hours of a falling tide. However, crude oil and fuel oil C penetrated into the sediments with much slower velocity than seawater and penetration depths were approximately 4.8 cm and 2.3 cm, respectively. A falling tide made the oils penetrate into the sediment but a rising tide did not work. By the combination of wave and tidal actions, two oils showed almost the same penetration behavior as the tidal fluctuation without wave. It is clear that tidal fluctuation plays major role for the penetration of oils into the sandy beach sediments.

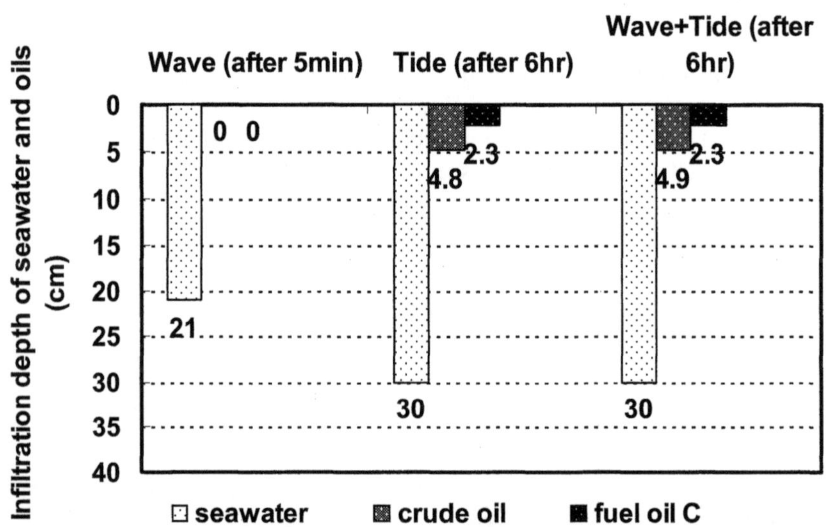

Fig. 4.2. Infiltration behaviors of seawater, crude oil and fuel oil C by wave and/or tidal actions

4.2.2 Volume of stranded oils

Fig.4.3 shows penetration depth of fuel oil C as affected by the volume of stranded oil. The penetration depths after the first tide were 2.5 cm and 6 cm for the stranded volume of 1 l m^{-2} and 4 l m^{-2}, respectively. Owens (1978) also reported that the depth and the rate of penetration depended on the viscosity and volume of the oil. It is clear that the penetration depths are significantly affected by the volume of stranded oil.

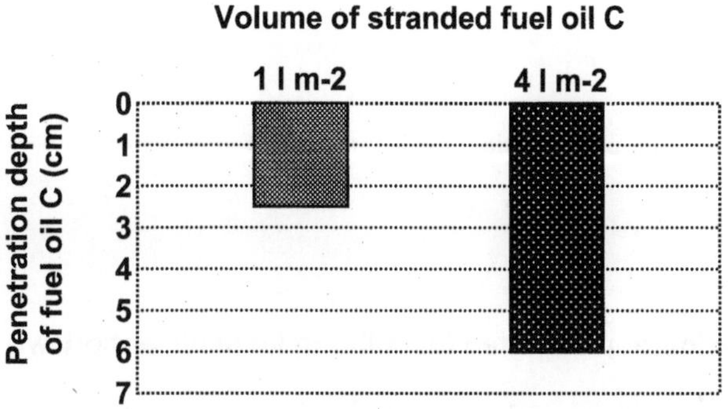

Fig. 4.3. Penetration depth of fuel oil C as affected by the stranded volume

4.2.3 Tidal Cycles and temperature

Penetration depths of fuel oil C over 15 tidal cycles observed in different water temperature conditions, 10 °C and 17 °C, are shown in Fig. 4.4. As seen from the observation window of the simulators, black area corresponds to the higher concentration of oils and gray area is the lower concentration. Significant penetration was noted in the first tidal cycle both in high and low water temperature. There was little penetration in the following tidal cycles. Almost no more penetration was noted for 15th tidal cycles in 10 °C, the black area lowered from 1.5 cm after the first tide to 2.0 cm in 15 tidal cycles. The higher temperature, 17°C, enhanced penetration. The black layer gradually lowered down to 3.0 cm in 15 tidal cycles. The similar trend was observed also for crude oil. Penetration was the largest in the first tide and the more penetration was noted in the higher temperature. Fig. 4.5 shows vertical distribution of hydrocarbon concentration in the sediments after the 15th tidal cycle. More than 80% of the oil was concentrated at the top 2 cm layer both in high and low water temperature. The higher temperature transferred the more oil to the lower layer. It is most likely that the penetration is highly dependent on the viscosity of oils affected by temperature.

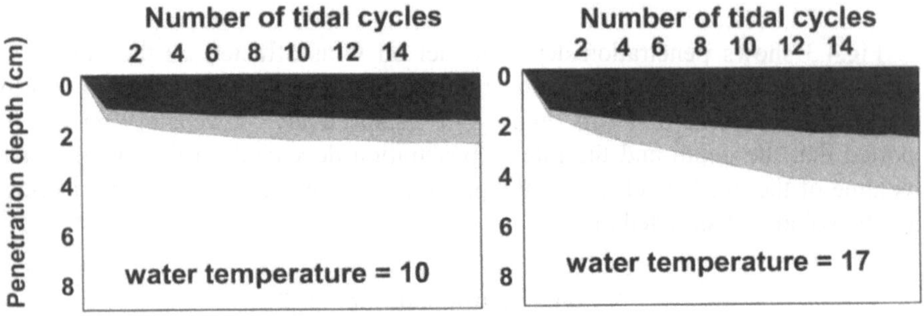

Fig. 4.4. Penetration of fuel oil C by tidal cycles under different water temperature conditions

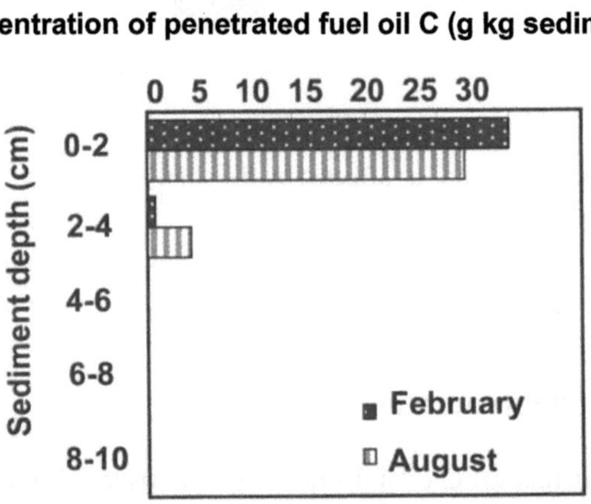

Fig. 4.5. Vertical concentration of penetrated fuel oil C after 15 tidal cycles (ND; not detected)

4.2.4 Viscosity of oil

The relationship between the viscosity of fuel oil C and temperature are shown in Fig. 4.6. Similar trend is also noted for crude oil. Viscosities of both oils are highly dependent of temperature and decreased with the increase in temperature. It must be noted that viscosity of fuel oil C is significantly higher than that of crude oil as shown in Table 4.1.

To confirm the effects of viscosity on oil penetration, penetration of silicon oil with different viscosity was studied in the same conditions as fuel oil C and crude oil. Fig. 4.7 shows the relationship between viscosity and penetration depth of silicon oils. At the lowest viscosity of silicon oil (12 $mm^2 s^{-1}$ at 15 ℃), the penetration depth was 15.3 cm, whereas the depth was only 1.1 cm at the highest one (21,000 $mm^2 s^{-1}$ at 15 ℃). It is clear that penetration depth increases with decrease in viscosity, i.e. oils with the lower viscosity penetrates down to the deeper sediments.

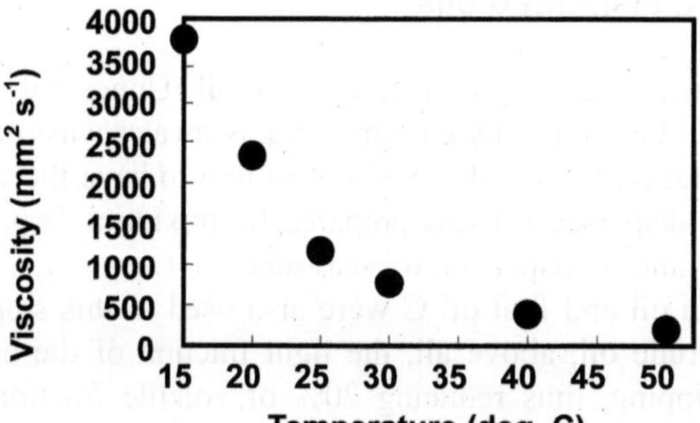

Fig. 4.6. Relationship between viscosity and temperature for fuel oil C

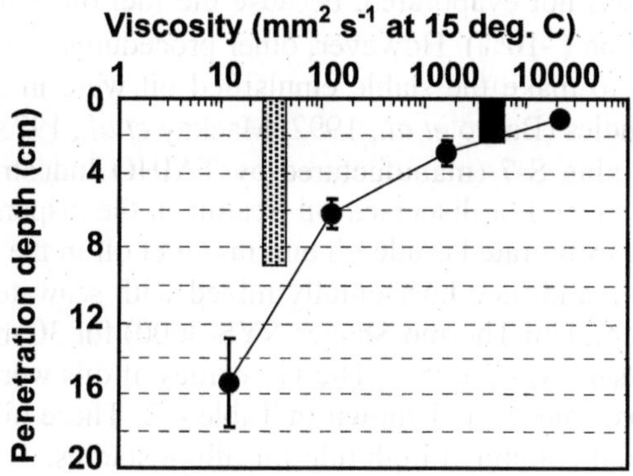

Fig. 4.7. Relationship between viscosity and penetration depth in silicon oil, crude oil (left bar) and crude oil (right bar)

The penetration depths of the crude oil and the fuel oil C were also shown in Fig. 4.7. The good agreement suggests that viscosity is one of the major factors to control the penetration depth of oils by tidal actions. The penetration depth of the fuel oil C was equal to that of silicon oil with equivalent viscosity. However, there was a little difference between crude oil and equivalent silicon oil. Reed *et al.* (1989) and Bear *et al.* (1991) reported that the penetration depth of spilled oil into the sandy beach sediments depended on the viscosity and specific gravity of the oil. It is most likely that the difference is probably due to the difference in specific gravity of oils. Specific gravity of crude oil is 0.87g cm^{-3} and that of silicon oil is 0.96 g cm^{-3} at 15° C. It must be noted that the specific gravity of the fuel oil C was almost same as that of the silicon on. Therefore, the penetration of oil into the sandy beach sediments is mainly controlled by the viscosity of oils.

4.2.5 Weathered and dispersed oils

The weathered oil used was prepared from crude oil (Upper Zakum). Volatile fractions were removed before use by gas stripping. Seawater was mixed with the evaporated oil by the ratio of 7:3 and mixed for 30 min to have the w/o emulsion of weathered oil. The dispersed oil was prepared by mixing a dispersant (Taiho Self Mixing S-7) by a ratio of dispersant:oil:seawater = 1:1:2.

Upper Zakum crude oil and fuel oil C were also used in this study. To make weathered oil using crude oil, above all, the light fraction of the crude oil was evaporated by air stripping, thus reducing 20% of volatile fraction by weight. Then the evaporated oil was emulsified with 80% (v v^{-1}) seawater by vertically vigorous shaking in a separatory funnel for 30 min equipped in shaker (TAITEC, Co., Recipro Shaker SR-Ⅱw) to obtain stable water-in-oil emulsions (w/o emulsions) formed when the oil is spilled at sea. To make weathered oil, on the other hand, fuel oil C was not evaporated, because the fuel oil C has lower content of volatile light fraction (~10%). However, other procedures were same as crude oil. These procedures to make the stable emulsified oil were in consideration of the previous oiling studies (Babro *et al.*, 1992; Mackay *et al.*, 1973).

TAIHO Self Mixing S-7 (manufactured by TAIHO Industries Co., LTD.) was selected as a dispersant. For dispersed-oil treatment, the dispersant was thoroughly premixed with the evaporated crude oil and raw fuel oil in the oil-to-dispersant ratio of 25:1 (w w^{-1}), and then horizontally mixed with seawater in a beaker set in shaker (EYELA Co., Uni Thermo Shaker NYS-3000) for 30 min. The ratio of water to oil to dispersant was 25:25:1. The viscosities of oils were determined by using brook-field viscometer, and shown in Table 4.2. These oils were applied into the sediment packed column at high tide for oiling studies. A typical test duration consisted of a further 16 tidal excursions, twice daily. Added oil volume was 1, 2, 4 and 8 L m^{-2} based on the previous studies on oil spills (Smith *et al.,* 1981; Delaune *et al.,* 1984; Little, 1987; Lin *et al.,* 1996; Oudot *et al.,* 1998).

Table 4.2. Viscosities of used oils in this study

Viscosity (cP) at 18℃	Normal	20% Evaporated	Oil +Dispersant	Oil +Dispersant +Water	Weathered
Crude oil	19	124	115	28	700
Fuel oil C	2720	-	1000	42	13100

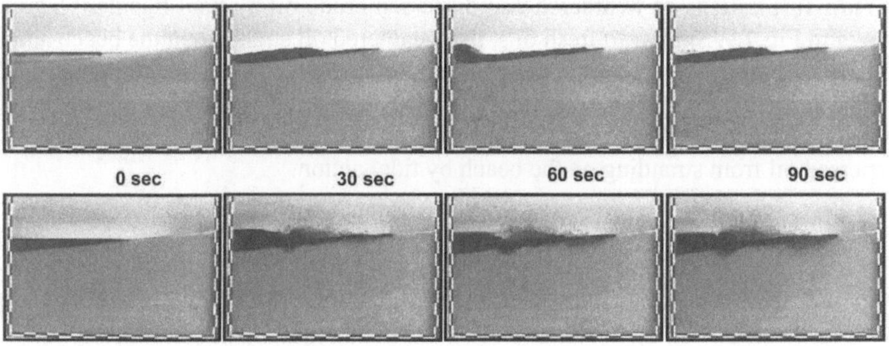

Fig. 4.8. Behavior of weathered oil (top) and dispersed oil (bottom) on sandy beach by wave actions

Wave. Fig. 4.8 shows behaviour of the weathered and dispersed crude oil near sandy beach by wave actions. Similar to fresh crude oil and fuel oil C (Fig.4.1), the weathered oil did not strand on the beach nor penetrate into the sediments by wave actions. This is because w/o emulsion still have low specific gravity. However, the dispersed oil penetrated into the sediments like seawater. The dispersant could disperse oil particles into seawater and changed physical characteristics of crude oil similar to seawater, whereas it must be noted that the penetration behavior is not the same as that of seawater. The periodic wave could not produce clear infiltration of the dispersed oil from the wave braking run-up point with radial direction as observed for seawater.

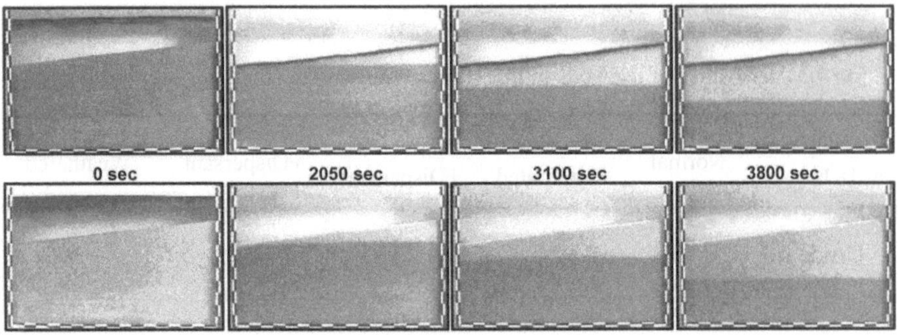

Fig. 4.9. Behavior of weathered oil (top) and dispersed oil (bottom) on sandy beach by tide

Tide. Infiltrations of weathered and dispersed crude oil by tidal fluctuation are shown in Fig.4.9. Similar to fresh oils, the weathered oil stranded on the beach and penetrated into the sediments. The dispersed oil, however, could not strand on the beach nor penetrate into the sediments. The dispersed oil has low interfacial tension by surfactant and still has low specific gravity. These factors prevented the dispersed oil from stranding on the beach by tidal action.

Wave+Tide. Infiltration of the weathered and the dispersed crude oil both by wave and tidal actions are shown in Fig.4.10. By the combination of waves and tides, both the weathered and dispersed oil penetrated into sandy beach sediments with the different driving forces, i.e. the weathered oil by tides and the dispersed oil by waves.

Fig. 4.10. Behavior of weathered oil (top) and dispersed oil (bottom) by wave and tidal actions

Fig. 4.11 shows penetration depths of the weathered and the dispersed oil in comparison with seawater and crude oil. The weathered oil did not infiltrate into the sediments by wave actions, whereas it penetrated into the sediments to 4.3 cm from the surface by 6 hours of falling tide and by the combination of wave and tidal actions. Although depths of penetration were smaller than those of crude oil, the weathered oil showed almost the same penetration behavior with that of crude oil. The dispersed oil penetrated into sediments to 12 cm by wave actions for 5 min. The depth was smaller than that of seawater. As shown in Fig. 4.9, tide did not penetrate dispersed oil into sediments. The penetration depth of the dispersed oil by the combination of waves and a falling tide for 6 hours was smaller than that of wave actions for 5 min. It is most likely that the falling tide removed the dispersed oil from the sediments because the specific gravity of the dispersed oil was low. Similar to fresh oils, tidal fluctuation played major role on the penetration of weathered oil. However, contrary to seawater, wave action was the major factor the penetration of dispersed oil.

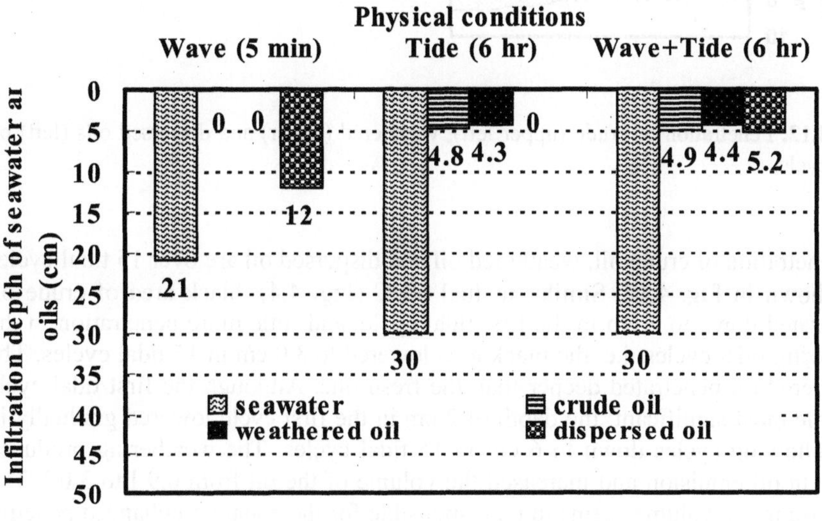

Fig. 4.11. Penetration depths of seawater and spilled oils in sandy beach sediments by wave and/or tidal actions

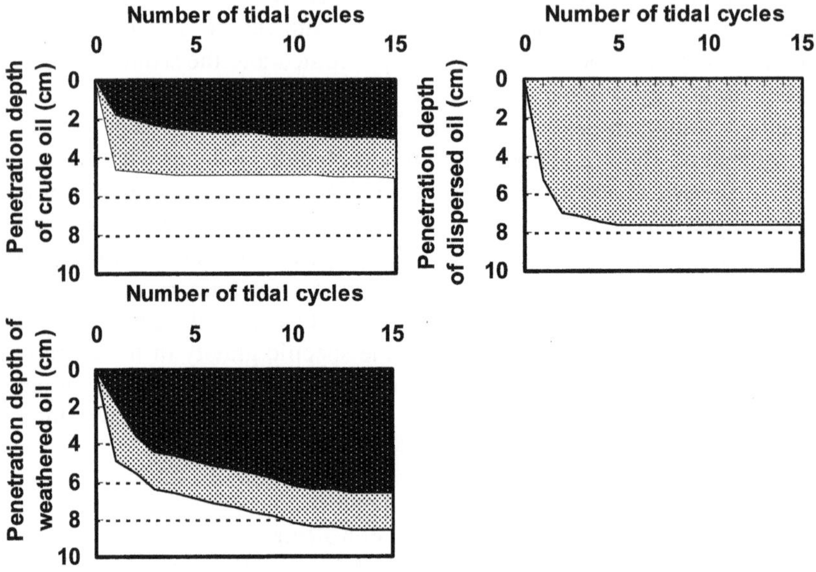

Fig. 4.12. Penetration of crude (upper left), weathered (upper) and dispersed oils (left) by tidal cycles

Penetration of crude oil, weathered oil and dispersed oil are over 15 tidal cycles are shown in Fig. 4.12. Similar to fuel oil C (Fig. 4.4), black area of crude oil penetrated down to 2 cm in the first tidal cycle, and little more penetration in the following tidal cycles, i.e. the black area lowered to 3.0 cm in 15 tidal cycles. The weathered oil penetrated deeper than the fresh one. Although the first tidal cycle was the most significant, the depth of 2 cm in the first cycle lowered gradually in the following cycles down to 6 cm in 15 tidal cycles. The weathering produced water in oil emulsion and increased the volume of the oil from 0.9 l to 3.0 l. The increase in the volume seems to be responsible for the apparent enhanced penetration irrespective of the increase in viscosity. The dispersed oil was solubilized into seawater and the amount of stranded oil was small. Therefore no black band was observed from the beginning. Most of the oil left the sediments for seawater in the first tide.

Fig. 4.13 shows vertical distribution of hydrocarbon concentration in the sediments both in the first and 15th tidal cycles. Almost all the weathered oil located at the top 2 cm layer, whereas little penetration was noted for the dispersed oil as shown in Fig. 4.12. The similar concentration from the surface to 6 cm for the weathered oil in 15th tidal cycle indicated gradual penetration of the oil. However, most of the crude oil remained at the surface layer even in the 15th cycle.

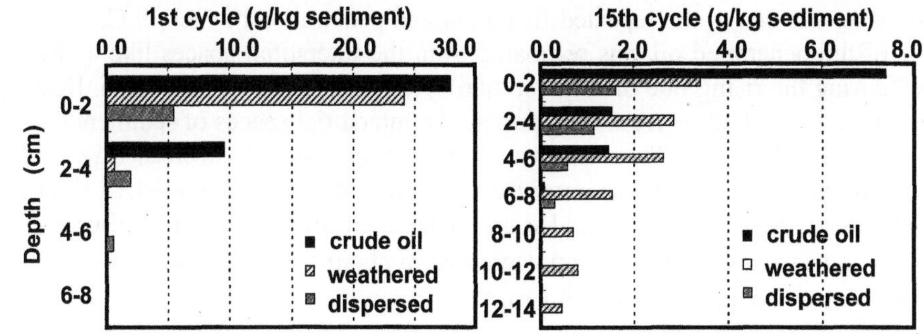

Fig. 4.13. Vertical concentration of penetrated crude, weathered and dispersed oil after the first and 15th tidal cycles

4.3 Effects of the penetrated oils into sandy beach sediments on seawater infiltration by waves

Figure 4.14 shows volume of seawater infiltrated (expressed by area) at the 1st and 4th tidal cycle into the sediments penetrated by fuel oil C. Breaking wave height and wave period were 30 mm and 0.8 sec, respectively. The amount of fuel oil C applied was 1 l • m^{-2}. After the first tidal cycle, the volume of seawater infiltration decreased to half comparing with the condition without oil. The volume of seawater through the penetrated oil layer in the 4th tidal cycle was far smaller than that of the first tidal cycle.

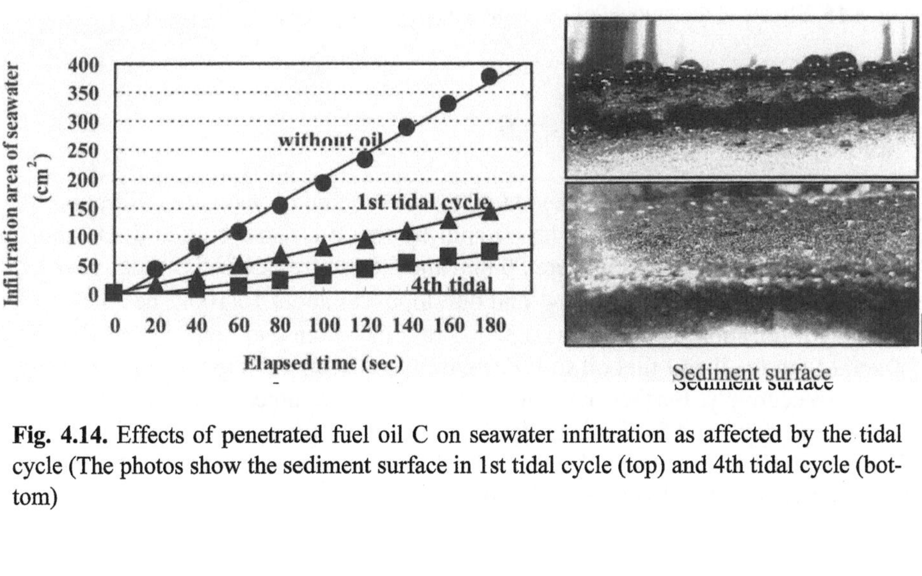

Fig. 4.14. Effects of penetrated fuel oil C on seawater infiltration as affected by the tidal cycle (The photos show the sediment surface in 1st tidal cycle (top) and 4th tidal cycle (bottom)

The difference in seawater infiltration between the first tidal cycle and the 4th tidal cycle was due to the repeated floatation and penetration of fuel oil C. About 20 % of the penetrated oil was re-floated from the interstitial spaces like a "balloon" during the rising tide of the first tidal cycle (see photos in Fig. 4.14). However, the re-floated oil penetrated again into the interstitial spaces of sediment during the next falling tide. The amount of re-floatation gradually decreased in tidal cycles and no re-floatation was observed in the 4th tidal cycle. Figure 4.15 shows the effects of the amount of stranded and penetrated oil on seawater infiltration. The amount of seawater infiltration decreased to about a half in case of 1 l • m^{-2}, while it almost stopped in case of 4 l • m^{-2}.

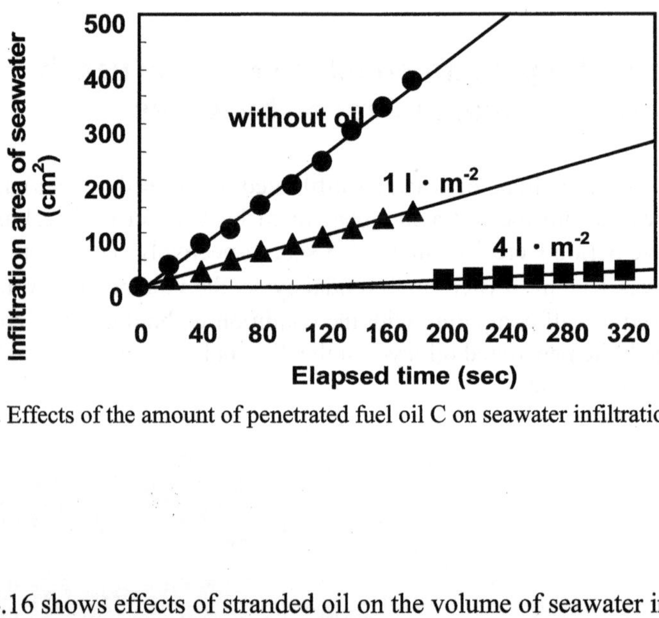

Fig. 4.15. Effects of the amount of penetrated fuel oil C on seawater infiltration

Fig. 4.16 shows effects of stranded oil on the volume of seawater infiltration by waves during tidal cycles after the oil spill by fuel oil C and crude oil. Breaking wave height and wave period were 50 mm and 0.8 sec, respectively. Tidal velocity was controlled at 0.009 cm • sec-1 and the slope was about 10/100. The amount of seawater infiltration by wave was 0.32 m3 m-2 day-1. In two tidal cycles after the strand of crude oil and fuel oil C, it decreased to 0.24 m^3 m^{-2} day^{-1} and 0.20 m^3 m^{-2} day^{-1} , respectively. Further but gradual decreases were noted in 10 tidal cycles to 0.22 m^3 m^{-2} day^{-1} and 0.17 m^3 m^{-2} day^{-1} , respectively for crude oil and fuel oil C. The penetrations of crude oil and fuel oil C decreased seawater infiltration by waves down to 1/3 and 1/2, respectively.

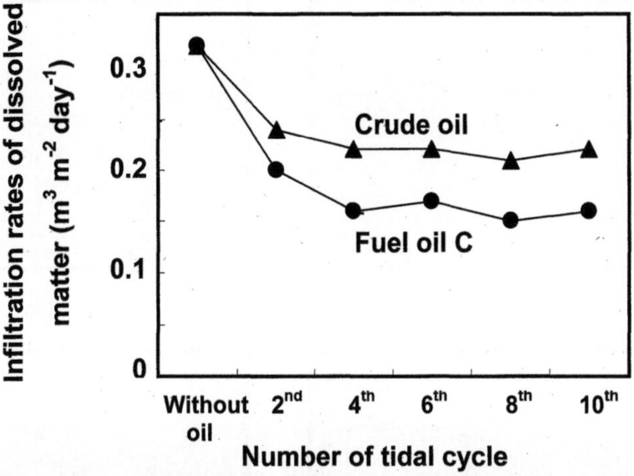

Fig. 4.16. Effects of penetrated crude oil on dissolved matter as affected by the tidal cycle

Fig. 4.17 shows vertical distributions of the penetrated oils in the 15th tidal cycles. Even after the 15th cycles, most of the stranded oils remained at the surface layer, i.e. 72 % and 98 % of crude oil and fuel oil C were in the 0-2 cm layer, respectively. The higher concentration of fuel oil C than crude oil at the surface and clogging of sediment pores may be responsible for the more significant decrease in the seawater infiltration by fuel oil C (Fig. 4.16). It is most probable that the decrease in seawater infiltration may cause deterioration of sandy beach ecosystem by the decrease in the supply of dissolved substances.

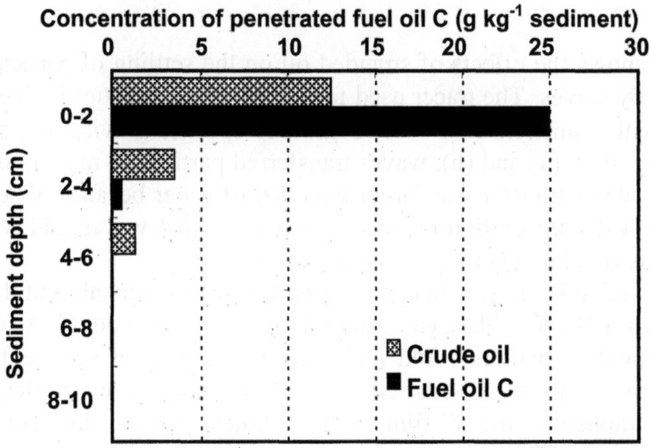

Fig. 4.17. Vertical distribution of penetrated oils in the 15th tidal cycle

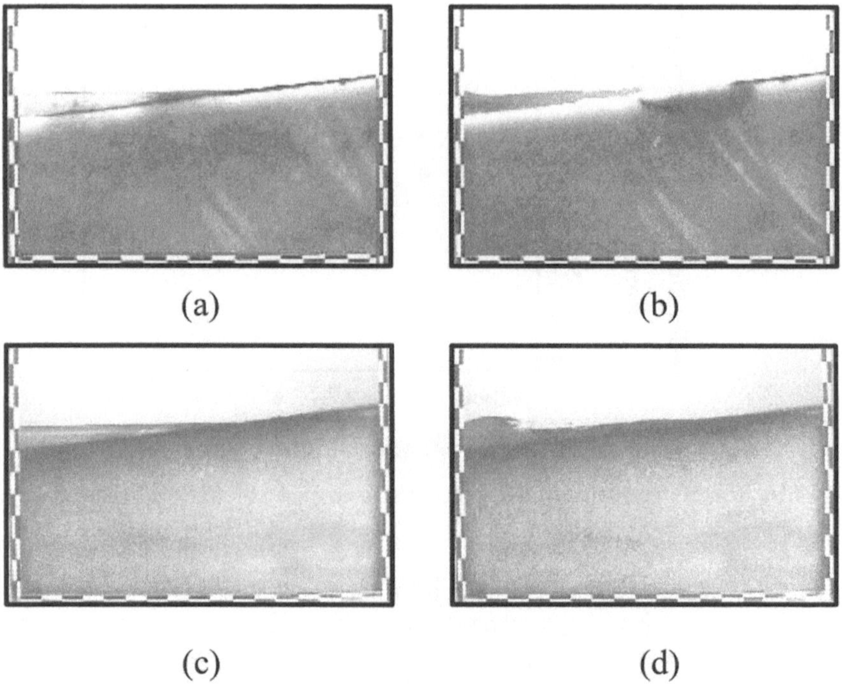

Fig. 4.18. Settling of particulate matters through the oiled (photo, c and d) and unoiled (photo; a and b) porous media. Elapsed times are 0 sec (photo; a and c) and 60 sec (photo; b and d)

Fig.4.18 shows the effects of stranded oil on the settling of particulate matters in seawater by waves. The tracer used for the particulate matter is fine particles of anthracite with diameters less than 75 μm and specific gravity of 1.5 g cm^{-3}. As shown in Fig. 4.18 (a) and (b), waves transferred particulate matters into the sediments. The rate of transfer was larger than that of water because of settling in addition to seawater infiltration by waves. It is clear that waves and sedimentation transported particulate matters into the sediments.

Fig. 4.18 (c) and (d) show behavior of particulate matters around the surface of the sediments in the 4th tidal cycle after oil pollution by crude oil. As shown in the figures, particulate matters could not settle into the oiled sediments. Fig. 4.19 shows enlarged photos of sediment particles (glass beads). Particulate matters seem to be trapped by the oil film of the sediment particles and could not settle into the sediments. Thus, oil pollution may prevent the supply of particulate matters like bacteria, planktons and silt to benthic organisms and deteriorate benthic ecosystem.

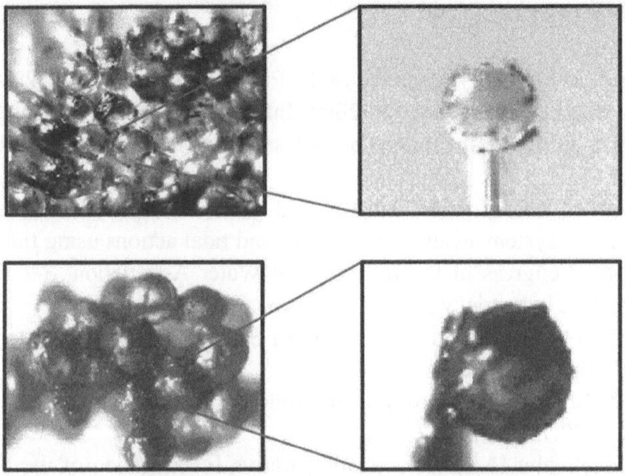

Fig. 4.19. Adhesion of particulate matters on the glass beads coated by crude oil (top) and fuel oil C (bottom). The diameter of glass bead is 1mm.

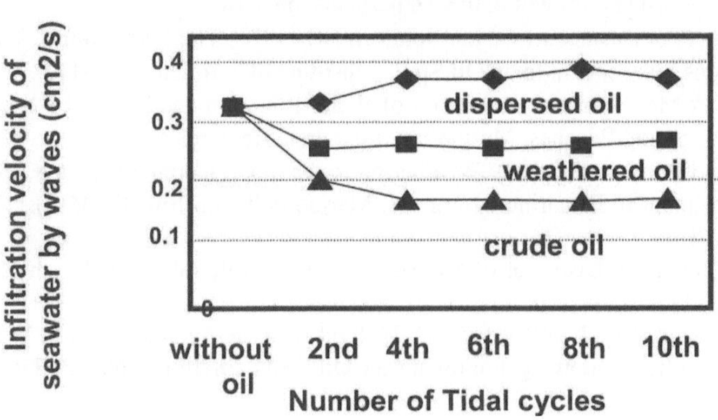

Fig. 4.20. Effects of stranded dispersed, weathered and crude oil on seawater infiltration as affected by the tide cycle

Fig. 4.20 shows effects of stranded dispersed, weathered and crude oils on seawater infiltration by waves as affected by tidal cycles. The amount of seawater infiltration by waves without oils was 0.32 $m^3 m^{-2} day^{-1}$. The decrease in the rate of seawater infiltration by the penetration of the weathered oil was smaller than that of crude oil. In two tidal cycles after the strand of oils, crude oil decreased the rate down to 0.20 $m^3 m^{-2} day^{-1}$, whereas it was 0.26 $m^3 m^{-2} day^{-1}$ in the case of weathered oil. No decrease was noted in the case of dispersed oil. As shown in Fig.4.13, most of the crude oil remained at the surface layer even in the 15th cycle, whereas little penetration was noted for the dispersed oil. Also, the smaller amount of the weathered oil penetration did not decrease seawater infiltration than crude oil.

References

Babro M and Fingas M (1992) When Oil Spills Emulsify, Chemtech, pp 236-241

Carman KR, Binanchi TS and Klope F (2000) Influence of grazing and nitrogen on benthic algal blooms in diesel fuel-contaminated saltmarsh sediments. Environ Sci Technol 34:107-111

Cheong CJ, Cho KJ, Nishijima W, Baba E, and Okada M (2000) Effects of spilled oil on the tidal flat ecosystem–evaluation of wave and tidal actions using tidal flat simulator. In 1st World Congress of the International Water Association, 3-7 July 2000, Paris France, Book 1, Industrial wastewater treatment, pp 469-475

Delaune RD (1984). Effect of Oil on Salt Marsh Biota: Methods for Restoration. Environmental Pollution 36:207-227

Jhonston R (1970). The Decomposition of Crude oil Residence in Sand Columns. J Mar Biol Ass UK 50:925-937

Lin Q and Mendelssohn IA (1996). A Comparative Investigation of the Effects of South Louisiana Crude Oil on the Vegetation of Fresh, Brackish and Salt Marshes. Marine Pollution Bulletin 32(2):202-209

Little DI (1987). The Physical Fate of Weathered Crude and Emulsified Fuel Oils as a Function of Intertidal Sedimentology. Fate and Effects of Oil in Marine Ecosystems. Martinus Nijhoff Publishers, Printed in the Netherlands, pp 3-18

Mackay GDM, Mclean AY, Betancourt OJ and Johnson BD (1973) The Formation of Water-in Oil Emulsions Sub sequent to an Oil Spill. J Institute of Petroleum 164-172

McLachlan A (1982). A Model for the Estimation of Water Filtration and Nutrient Regeneration by Exposed Sandy Beaches. Marine Environmental Research 6:37-47

Oudot J, Merlin FX and Pinvidic P (1998). Weathering Rates of Oil Components in a Bioremediation Experiment in Estuarine Sediments. Marine Pollution Bulletin 45(2):113-125

Owens EH (1978). Mechanical Dispersal of Oil Stranded in the Littoral Zone. J Fish Res Board Can 35:563-572

Smith CJ, Delaune RD and Patrick WH (1981). A Method for Determining Stress in Wetland Plant Communities Following Following an Oil Spill. Environmental Pollution 26:297-304

Swannell PJ, Lee K and Mcdonagh M (1996). Field Evaluations of Marine Oil Spill Bioremediation. Microbiological Reviews 60(2):342-365

Vandermeulen JH, Buckley DE, Levy EM, Long BFN, McLaren P, and Wells PG (1979) Sediment penetration of *Amoco Cadiz* oil, potential for future release, and toxicity. Marine Pollution Bulletin 10:222-227

Wolfe DA, Hameedi M, Galt JJA, Watabayashi G, Short J, O' Clair C, Rice S, Michel J, Payne JR, Braddock J, Hanna S and Sale D (1994). The Fate of the Oil Spilled from the Exxon Valdez. Environmental Science Technology 28:561A-568A

5 Theoretical study of oil pollution

5.1 Introduction

We discussed effects of waves on a tidal flat ecosystem in Chap. 3 and understood remarkable works of nature in the neighborhood of the margin. In this chapter, we analyze what effects the intervention of oil gives to the harmonious relation between waves and tidal flat ecosystems.

Experimental observations of effects of spilled oil on seawater infiltration can be summarized.

1 Oil on a seabed under wave and tidal action

Spilled oil penetrates into the tidal flat due to tidal motion and not due to wave action. Repeated cycles of tidal motion harden the spilled oil layer over the surface of the tidal flat and result in decreased seawater infiltration into the tidal flat.

2 Effect of spilled oil on the infiltration of seawater

Prevention of fresh seawater infiltration will result in decreased supply of oxygen, nutrients, silts, and organic matter needed for survival of macrobenthic organisms in tidal flats.

Through theoretical methodology borrowed from the mathematical sciences, we observed the following by computer analysis:

(1) Process of drifting ashore of spilled oil accompanying waves.
(2) Penetrating process of oil into tidal flats and sea beds.
(3) Blocking effect of oil on seawater infiltration.

5.2 Behavior of spilled oil in the surfzone

We have to remodel relevant phenomena as a three-phase flow by adding the oil phase to air and water phases. Further, oil effects are treated as interfacial interactions between the oil and seawater phases. We noted the following points in the process of building a mathematical model.

(1) Three-phase free surface flow consists of air, water, and oil in the surfzone.

(2) Generally speaking, fuel oil C can be treated as a Newtonian fluid in a normal situation. However, if it spills out into the sea, it constitutes an emulsion as a consequence of mixing with seawater and/or volatilization of its smaller molecular weight components. Finally, it reaches a state resembling mousse floating on the sea surface, which may be treated by the Cross model.
(3) Oil has the property of adherence to sand. To describe this property mathematically, a friction-type boundary condition is introduced as an interfacial interaction between seawater and oil in sand.
(4) The change of water head in a tidal flat due to tidal motion.

5.3 Mathematical model

5.3.1 Notations and geometry

x_i	: Cartesian coordinates $(i = 1, 2, 3)$
u_i	: Velocity vector $(i = 1, 2, 3)$
u_n	: Normal component of velocity on the boundary
u_T	: Tangential component of velocity on the boundary
p	: Pressure
ρ	: Density
t	: Time
D_{ij}	: Rate of strain tensor
Ω_α	: Domain for α-phase, where $\alpha = \{air,\ water,\ oil\}$
$\Omega = \bigcup_\alpha \overline{\Omega_\alpha}$	: Total domain
Ω_L	: Domain occupied with sand
Ω_U	: $\Omega \backslash \overline{\Omega_L}$
χ_α	: Characteristic function representing a domain Ω_α, i.e., $\chi_\alpha = 1$ in $\overline{\Omega_\alpha}$ and $\chi_\alpha = 0$ in $\Omega \backslash \overline{\Omega_\alpha}$
χ_L	: Characteristic function representing a domain Ω_L
χ_U	: Characteristic function representing a domain Ω_U
χ_α^L	: $\chi_L \cdot \chi_\alpha$
χ_α^U	: $\chi_U \cdot \chi_\alpha$
ν^α	: Kinematic viscosity of α-phase
μ^α	: Viscosity of α-phase
σ_{ij}^α	: Stress tensor of α-phase
ρ^α	: Density of α-phase
c	: Coefficient of resistance force received from sand
g_T	: Friction coefficient between oil and water in sand
K_i	: ith component of external force
$\lvert A \rvert$	: Euclidean norm of A
Γ_B	: Boundary between Ω_{water} and Ω_{air}

Γ_I : Inflow boundary
Γ_S : Slip boundary
Γ_N : No-slip boundary
Γ_A : Outflow boundary
Γ_{OS} : Boundary of oil in sand, i.e., $\Gamma_{OS} = \partial\Omega_{oil} \cup \Omega_L$.

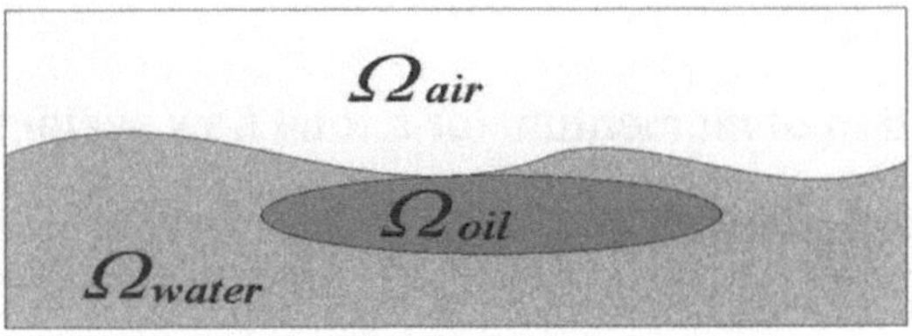

Fig. 5.1. Three-phase flow

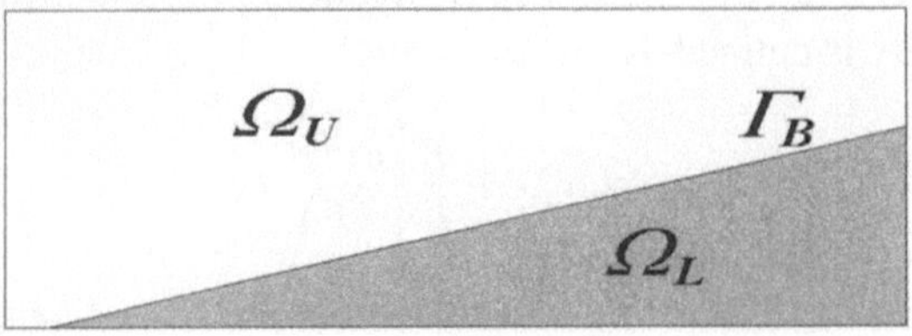

Fig. 5.2. Sloping beach

Figure 5.1 shows Ω_α where $\alpha = \{air,\ oil,\ water\}$ and Fig. 5.2 shows Ω_L and Ω_U. In this problem, the domain Ω is consists of six parts.

(1) The first part ($\Omega_{air} \cap \Omega_U$) is occupied with air.
(2) The second part ($\Omega_{water} \cap \Omega_U$) is filled with water.
(3) The third part ($\Omega_{water} \cap \Omega_L$) is occupied with water penetrated into sand from the sloping beach.
(4) The fourth part ($\Omega_{air} \cap \Omega_L$) is the dry part of sand filled with air.
(5) The fifth part ($\Omega_{oil} \cap \Omega_U$) is occupied by oil in seawater.
(6) The sixth part ($\Omega_{oil} \cap \Omega_L$) is occupied by oil penetrated into sand.

5.3.2 Conservation of mass for a total flow system

By repeating the same arguments as in Chap. 3, we have incompressibility conditions for fluid in each phase

$$\frac{\partial u_j^l}{\partial x_j} = 0, \text{ in } \Omega_{(l)}, \text{ for each } l, \ t > 0, \tag{5.1}$$

and the interface motion equation;

$$\frac{\partial \chi_l}{\partial t} + u_j^l \cdot \frac{\partial \chi_l}{\partial x_j} = 0, \ \text{in} \Omega_{(l)}, \text{ for each } l, \ t > 0. \tag{5.2}$$

Here, l means $\{water, air, oil\}$.

5.3.3 Conservation of momentum for a total flow system

Equations of motion for air and water These are introduced in the same way as in Chap. 3.

Equations of motion for spilled oil As mentioned in the Introduction, spilled oil achieves to a mousse-like state as a consequence of mixing with seawater. Spilled oil in such a state behaves as a non-Newtonian fluid. Various types of constitutive equations for non-Newtonian fluids have been proposed: (Bingham 1922; Cross 1965; Sisko 1958) We adopts the Cross model among them, in which a shear-dependent viscosity is defined as

$$\mu_{oil} = \mu_{oil}^{\infty} + \frac{\mu_{oil}^{0} - \mu_{oil}^{\infty}}{1 + (K\dot{r})^{m}}, \tag{5.3}$$

where $\dot{r}$ is a shear rate, μ_{oil}^0 is the limit of viscosity as $\dot{r} \to 0$, and μ_{oil}^{∞} is the limit as $\dot{r} \to \infty$. K and m are positive parameters. It should be noted that the Cross model reduces to the *power-law model* or the *Bingham model* by range restriction and redefinition of parameters. See (Barnes et al. 1989) for details. If one considers the range $\mu_{oil}^{\infty} \ll \mu_{oil} \ll \mu_{oil}^{0}$, the Cross model (5.3) reduces to

$$\mu_{oil} = K_1 \dot{r}^{n-1}, \tag{5.4}$$

where $K_1 = \dfrac{\mu_{oil}^0}{K^m}$ and $n = 1 - m$. Equation (5.4) is the well-known *power-law model* and n is called the power-law index.

Also, considering that $\mu_{oil} \ll \mu_{oil}^0$ and taking $m = 1$ yields

$$\mu_{oil} = \mu_{oil}^{\infty} + \frac{\sigma_y}{\dot{r}}, \tag{5.5}$$

where $\sigma_y = \dfrac{\mu_{oil}^0}{K}$. This relation can be rewritten as

$$\sigma = \mu_{oil}\dot{r} = \mu_{oil}^{\infty}\dot{r} + \sigma_y. \tag{5.6}$$

This is called the *Bingham model* with *yield stress* σ_y, at which the fluid begins to deform. Since the range of shear rate which should be covered in our problem is very wide, the Cross model was chosen.

Remark: *Dealing with a Bingham fluid poses the numerical problem of dividing by zero at the second term of the right-hand side of (5.5) when the fluid is at rest. This difficulty corresponds to the fact that* $\mu_{oil} \to \infty$ *as* $\dot{r} \to 0$ *in the Bingham model. A regularization technique of adding a small parameter to the denominator corresponds to assuming existence of a finite* μ^0_{oil} *as done in the Cross model.*

In this study, the following quantity is chosen as the shear rate:

$$D_{\Pi} = \frac{1}{2} \sum_{i,j} D_{ij} D_{ij}. \tag{5.7}$$

D_{Π} is one invariant of the rate of strain tensor which is invariant for coordinate transformations. Therefore, stress σ_{ij} of spilled oil is written as

$$\sigma_{ij} = -p\delta_{ij} + \left\{ \mu^{\infty}_{oil} + \frac{\mu^0_{oil} - \mu^{\infty}_{oil}}{1 + (KD_{\Pi})^m} \right\} D_{ij}. \tag{5.8}$$

Additionally, sand is treated as porous medium in which a resistance force proportional to fluid velocity acts on the fluid. Therefore, equations of motion for spilled oil in and out of sand are

$$\rho_{oil} \left(\frac{\partial u_i}{\partial t} + u_j \frac{\partial u_i}{\partial x_j} \right) = -\frac{\partial p}{\partial x_i} + \rho_{oil} K_i$$
$$+ \frac{\partial}{\partial x_j} \left\{ \left(\mu^{\infty}_{oil} + \frac{\mu^0_{oil} - \mu^{\infty}_{oil}}{1 + (KD_{\Pi})^m} \right) \left(\frac{\partial u_i}{\partial x_j} + \frac{\partial u_j}{\partial x_i} \right) \right\}, \tag{5.9}$$

and

$$\rho_{oil} \left(\frac{\partial u_i}{\partial t} + u_j \frac{\partial u_i}{\partial x_j} \right) = -\frac{\partial p}{\partial x_i} + \rho_{oil} K_i - \rho_{oil} c u_i$$
$$+ \frac{\partial}{\partial x_j} \left\{ \left(\mu^{\infty}_{oil} + \frac{\mu^0_{oil} - \mu^{\infty}_{oil}}{1 + (KD_{\Pi})^m} \right) \left(\frac{\partial u_i}{\partial x_j} + \frac{\partial u_j}{\partial x_i} \right) \right\}, \tag{5.10}$$

where c is proportional to the inverse of sand porosity.

5.3.4 Unified equations of motion for a three-phase flow

Conservation of momentum for the total system of three-phase flow is represented by use of motion equations for each flow.

$$\rho \left(\frac{\partial u_i}{\partial t} + u_j \frac{\partial u_i}{\partial x_j} \right) =$$
$$-\frac{\partial p}{\partial x_i} - c\chi_L u_i + \rho K_i + \frac{\partial}{\partial x_j} (\mu D_{ij}) \ \text{ in } \Omega, \ t > 0, \tag{5.11}$$

where

$$\mu = \mu_{air}\chi_{air} + \mu_{water}\chi_{water} + \left(\mu^{\infty}_{oil} + \frac{\mu^0_{oil} - \mu^{\infty}_{oil}}{1 + (KD_{\Pi})^m} \right) \chi_{oil}. \tag{5.12}$$

5.3.5 Adhesive phenomena of oil on a sandy beach

As an interfacial interaction, we formulate adhesion and sliding phenomena of oil in sand on the same way as done for $[\sigma_n]$ in Chap. 3 on the basis of a simplified Coulomb law for friction. Let $[\sigma_T]$ be the jump of tangential stress defined on the boundary between oil and water in sand:

$$\begin{cases} |[\sigma_T]| \leq g_T, \\ (|[\sigma_T]| - g_T) \cdot u_T = 0, \quad \text{on } \Gamma_{OS}. \\ [\sigma_T] \cdot u_T \leq 0, \end{cases} \tag{5.13}$$

where g_T is the friction coefficient. This means

$$\begin{cases} |[\sigma_T]| < g_T \longmapsto u_T = 0 & \text{(Adhesion)}, \\ |[\sigma_T]| = g_T \longmapsto u_T = 0 \text{ or } u_T \neq 0 & \text{(Sliding)}. \end{cases} \tag{5.14}$$

(5.13) is modified to

$$\begin{cases} |[\sigma_T]| \leq g_T, \\ g_T \cdot |u_T| + [\sigma_T] \cdot u_T = 0, \end{cases} \quad \text{on } \Gamma_{OS}, \tag{5.15}$$

and expressed as

$$-[\sigma_T] = g_T \cdot \partial(|u_T|) \qquad \text{on } \Gamma_{OS}, \tag{5.16}$$

by use of the definition of subdifferential.

In order to avoid difficulty due to singularity arising in numerical treatment of (5.16), Eq. (5.16) is regularized as

$$-[\sigma_{T_j}] = g_T \cdot \frac{u_{T_j}}{\sqrt{|u_T|^2 + \varepsilon_g^2}} \quad \text{on } \Gamma_{OS}, \ (j = 1, 2), \tag{5.17}$$

where ε_g is a small positive parameter.

5.4 Unified model for three-phase flow with adhesion of oil on a sandy beach

The incompressibility condition for a three-phase flow system is defined as

$$\frac{\partial u_j}{\partial x_j} = 0, \quad \text{in } \Omega, \quad t > 0. \tag{5.18}$$

Conservation of momentum for a total flow system with interfacial interactions is set as a single equation in the following way (Fujita et al. 1995):

$$\rho\left(\frac{\partial u_i}{\partial t}+u_j\frac{\partial u_i}{\partial x_j}\right)=-\frac{\partial p}{\partial x_i}-c\chi_L u_i$$
$$+\rho K_i+\frac{\partial}{\partial x_j}\left\{\mu\left(\frac{\partial u_i}{\partial x_j}+\frac{\partial u_j}{\partial x_i}\right)\right\}$$
$$+\rho\chi_L g_T\left\{\frac{u_{T_1}}{\sqrt{|u_T|^2+\varepsilon_g^2}}(\mathcal{T}_1\chi_{oil})_i+\frac{u_{T_2}}{\sqrt{|u_T|^2+\varepsilon_g^2}}(\mathcal{T}_2\chi_{oil})_i\right\}$$
$$\text{in } \Omega,\ t>0. \tag{5.19}$$

where

$$\mu(x_i,t)=\mu_{air}\chi_{air}(x_i,t)+\mu_{water}\chi_{water}(x_i,t)+\left(\mu_{oil}^{\infty}+\frac{\mu_{oil}^{0}-\mu_{oil}^{\infty}}{1+(KD_{\Pi}(x_i,t))^m}\right)\chi_{oil}(x_i,t). \tag{5.20}$$

The last term of the right hand side of (5.19) represents adhesion and sliding phenomena occurring at the interface between oil and sand. Tangential vectors along the boundary of domain Ω_{oil} are represented by $\mathcal{T}_1\chi_{oil}$ and $\mathcal{T}_2\chi_{oil}$ in (5.19) and defined as follows.

$$\mathcal{T}_1\chi_{oil}=\left(\frac{\partial\chi_{oil}}{\partial x_2},\ -\frac{\partial\chi_{oil}}{\partial x_1},\ 0\right), \tag{5.21}$$

$$\mathcal{T}_2\chi_{oil}=\left(\frac{\partial\chi_{oil}}{\partial x_1}\frac{\partial\chi_{oil}}{\partial x_3},\ \frac{\partial\chi_{oil}}{\partial x_2}\frac{\partial\chi_{oil}}{\partial x_3},\ \left(\frac{\partial\chi_{oil}}{\partial x_3}\right)^2-|\nabla\chi_{oil}|^2\right). \tag{5.22}$$

In fact, $\mathcal{T}_i\chi_{oil}$ is a delta-measure supported on Γ_{OS} through which the friction condition (5.17) is included in (5.19).

The phase equation for characteristic functions is

$$\chi_{air}+\chi_{water}+\chi_{oil}=1 \ \text{ in } \Omega,\ t>0. \tag{5.23}$$

Motion equations of free surfaces for air and oil are

$$\frac{\partial\chi_{air}}{\partial t}+u_j\frac{\partial\chi_{air}}{\partial x_j}=0 \ \text{ in } \Omega,\ t>0, \tag{5.24}$$

$$\frac{\partial\chi_{oil}}{\partial t}+u_j\frac{\partial\chi_{oil}}{\partial x_j}=0 \ \text{ in } \Omega,\ t>0. \tag{5.25}$$

Finally, outer boundary conditions are prescribed as

$$\begin{cases} u_n=u_0\sin\omega t,\\ u_T=0,\end{cases} \quad \text{on } \Gamma_I(t), \tag{5.26}$$

$$\begin{cases} u_n=0,\\ \dfrac{\partial u_T}{\partial n}=0,\end{cases} \quad \text{on } \Gamma_S, \tag{5.27}$$

$$\begin{cases} u_n = 0, \\ u_T = 0, \end{cases} \qquad \text{on } \Gamma_N, \tag{5.28}$$

$$\begin{cases} \sigma_n = 0, \\ u_T = 0, \end{cases} \qquad \text{on } \Gamma_A, \tag{5.29}$$

where ω and u_0 are an angular velocity and an amplitude of an incoming wave, respectively. We intend to clarify physical properties of spilled oil by seeking, numerically, the unknowns $\{u_i, p, \chi_\alpha\}$ for the unified model problem.

5.5 Numerical results

5.5.1 Process of drifting ashore of spilled oil

We observe from numerical simulation that (1) spilled oil placed in the offing starts to move along with wave motion; (2) some part of oil fluidizes gradually, broadens thinly on the sea surface, drifts ashore, and sticks to the sand beach due to interaction of a friction type; (3) on the other hand, the other part of oil which does not stick to sand is torn off under influence of seawater strain and comes back to the offing.

When spilled oil penetrates into the sand, the area of the contact surface between oil and sea water lessens. Then it becomes difficult for the strain energy to propagate from sea water into oil through the contact surface; that causes increased kinematic viscosity of oil which in turn aids progress of oil solidization in sand.

Fig. 5.3 shows motion of spilled oil due to wave breaking action. We can see that the mass of the spilled oil separates into two parts. One part adheres to the sandy beach and the other part flows out. See the appendix which includes color figures.

Fig. 5.4 shows numerical result of 3D case. Some results in this book including Fig. 5.4 were computed by use of a numerical method due to fictitious domain method, whereas a comprehensive numerical method to deal with the probrem is proposed in this book. It is known that the fictitious domain method can solve effectively various free surface probrems. See (Kawarada and Suito 1997) for details. It also should be noted that there is no essential difference between them through numerical results.

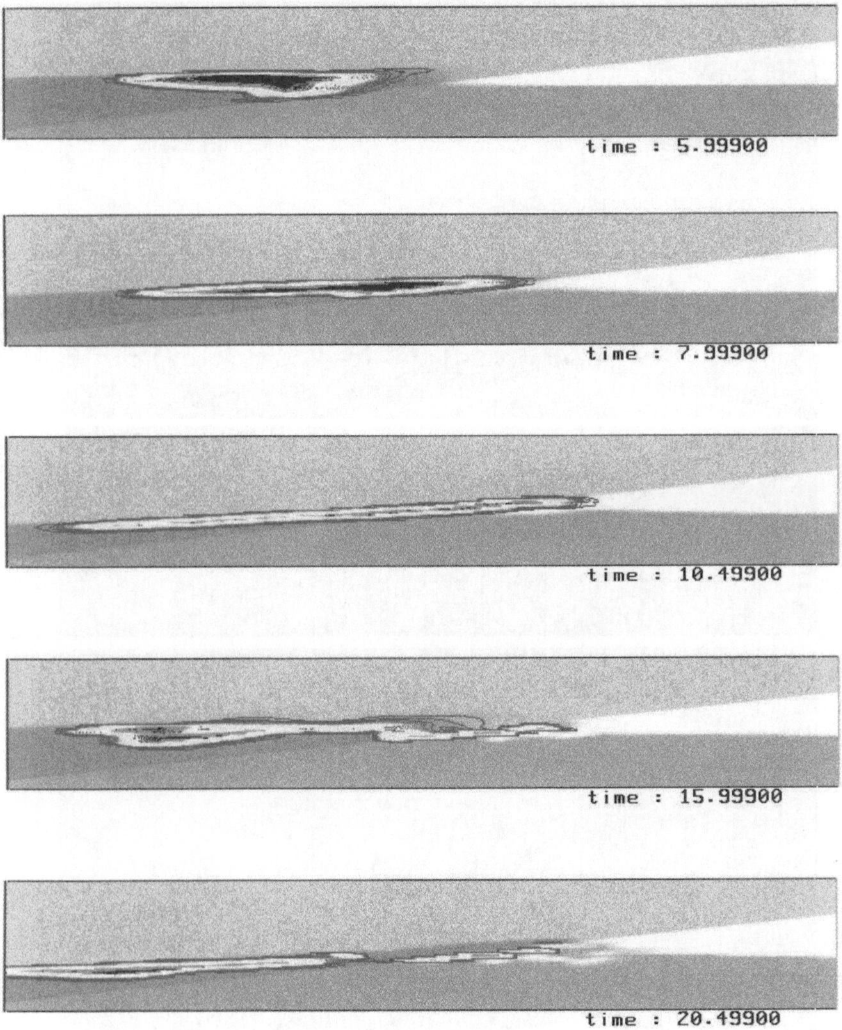

Fig. 5.3. Motion of spilled oil due to wave breaking action (2D case)

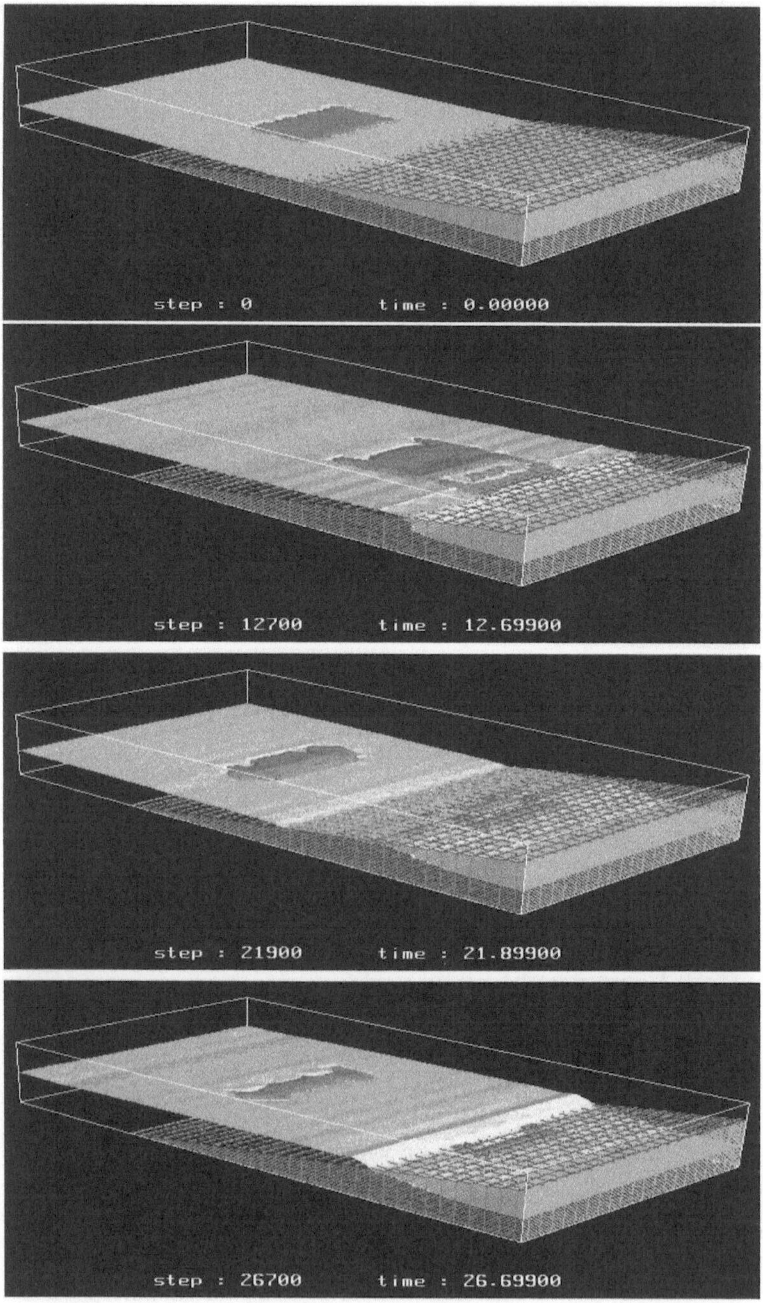

Fig. 5.4. Motion of spilled oil due to wave breaking action (3D case)

5.5.2 Deformation of oil

Figures 5.5, 5.6, and 5.7 show deformation processes where μ_{oil}^{0} are 10^4, 10^3, and 10^2 Pa·sec, respectively. The value for μ_{oil}^{∞} is set at 10^{-1} in all cases, which is the viscosity of fuel oil C in a normal situation. The parameters K and m in the Cross model are set at 1.0 and 0.95, respectively. In these figures, brown, yellow, dark-blue, and light-blue colors indicate $\Omega_{water}(t) \cap \Omega_L$ (seawater in sand), $\Omega_{air}(t) \cap \Omega_L$ (air in sand), $\Omega_{water}(t) \cap \Omega_U$ (seawater out of sand), and $\Omega_{air}(t) \cap \Omega_U$ (air out of sand), respectively. See color figures corresponding to these figures included in the appendix. Spilled oil (Ω_{oil}) is indicated by a violet color. Top figures show that the initial shapes of spilled oil are rectangular.

In the case of $\mu_{oil}^{0} = 10^4$ (Fig. 5.5), spilled oil acts as if it were a solid body. Evidently, spilled oil cannot be deformed easily though wave motion acts on it in the sequence of figures. In the case of $\mu_{oil}^{0} = 10^3$ (Fig. 5.6), spilled oil is slightly softened. It is stretched by receding waves, but is not torn off. On the other hand, in the case of $\mu_{oil}^{0} = 10^2$ (Fig. 5.7), spilled oil is easily deformed. It can be seen that the spilled oil is highly stretched, torn off, whereupon it penetrates into sand by repetition of wave motion.

5.5.3 Adhesion and sliding phenomena between oil and water in a sandy beach

Figures 5.8 and 5.9 are magnified images around the oil showing adhesion phenomena of spilled oil on a sand beach. White arrows are velocity vectors. In the case of $g_T = 0.0$ (Fig. 5.8), oil touches the sand beach, but does not adhere to it. On the other hand, in the case of $g_T = 100.0$ (Fig. 5.9), some of the oil adheres to the sand beach and is torn off. Then the divided portions are united again, but part of the oil still adheres to the sand beach. It shows the effect of a friction type boundary condition imposed on the boundary between oil and water in sand.

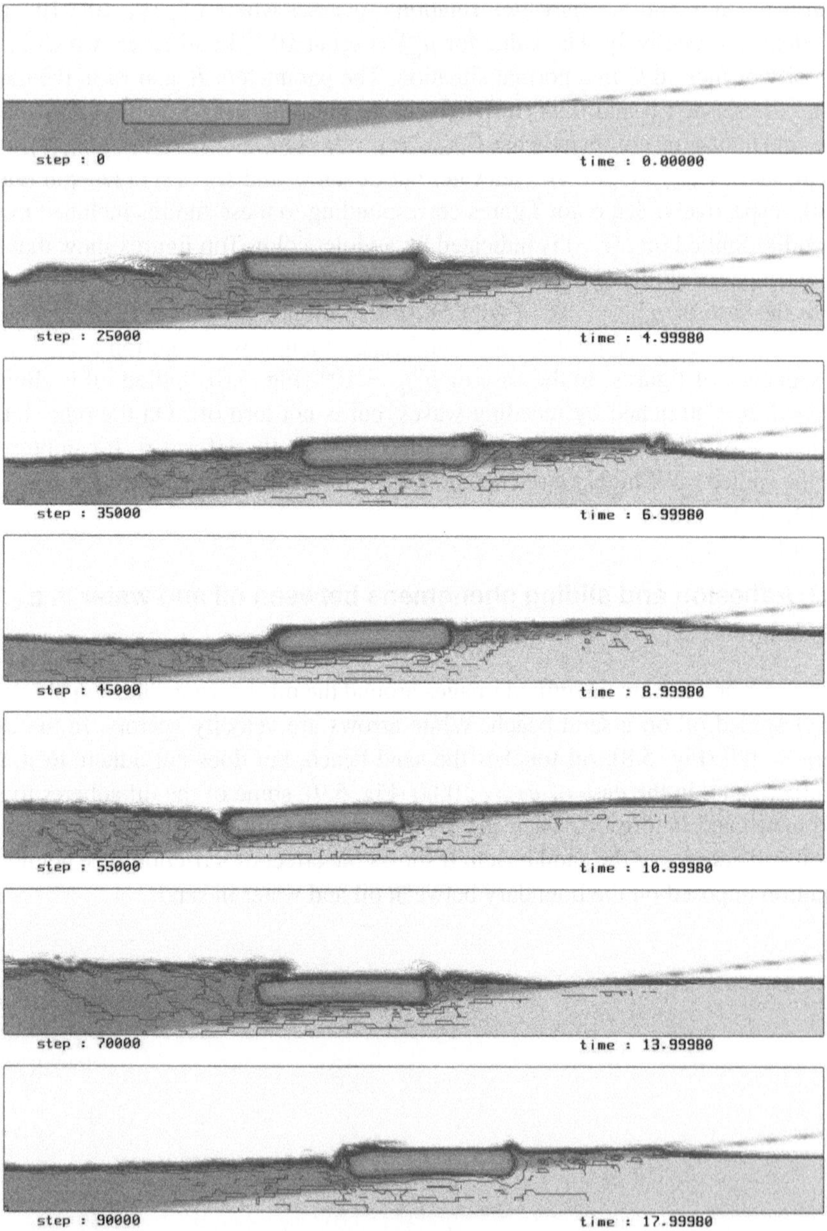

Fig. 5.5. Oil deformation ($\mu^0_{oil} = 10^4$)

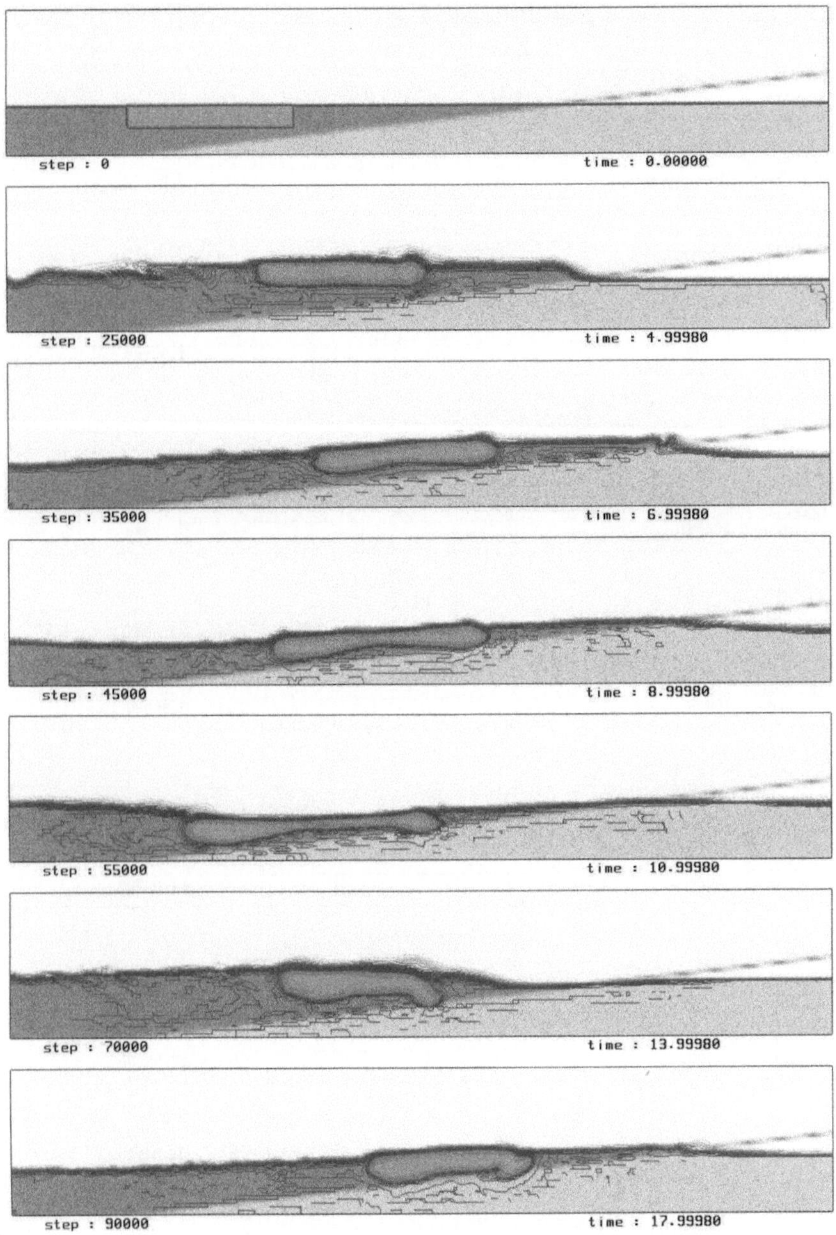

Fig. 5.6. Oil deformation ($\mu^0_{oil} = 10^3$)

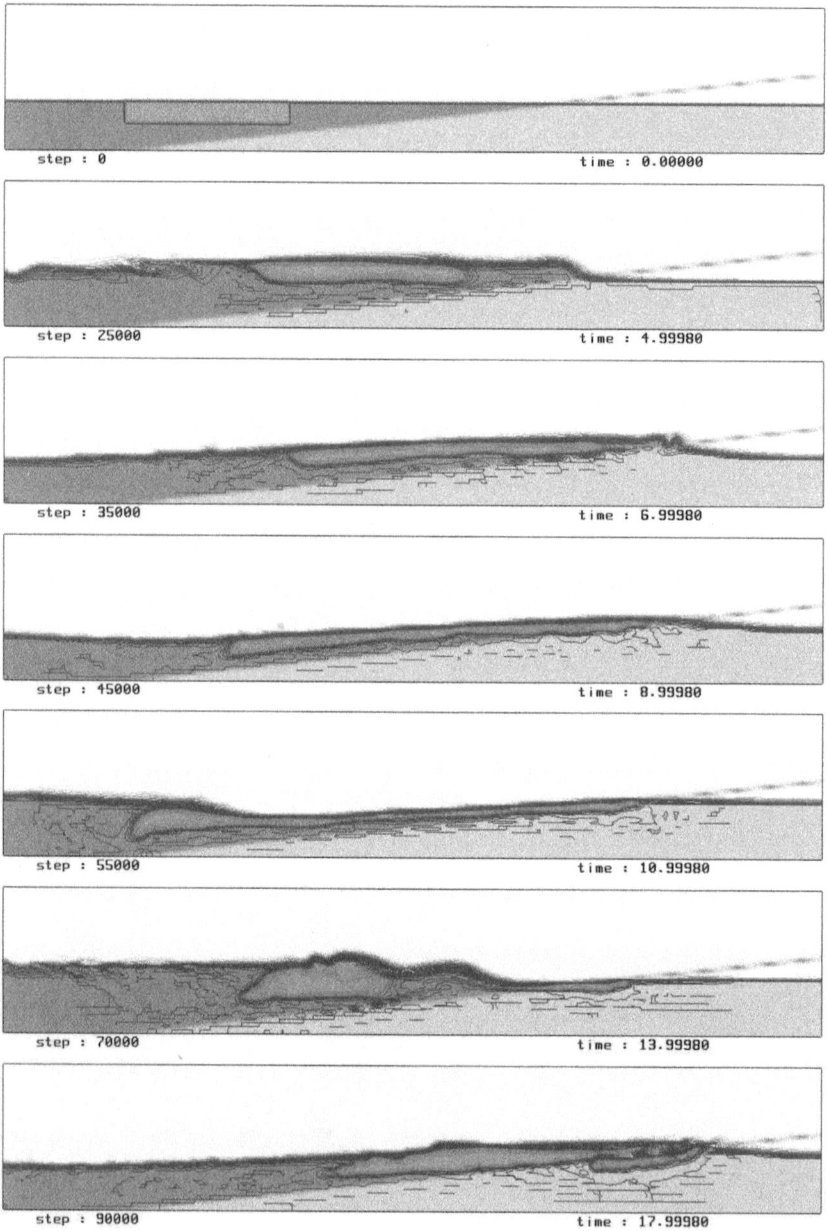

Fig. 5.7. Oil deformation ($\mu_{oil}^0 = 10^2$)

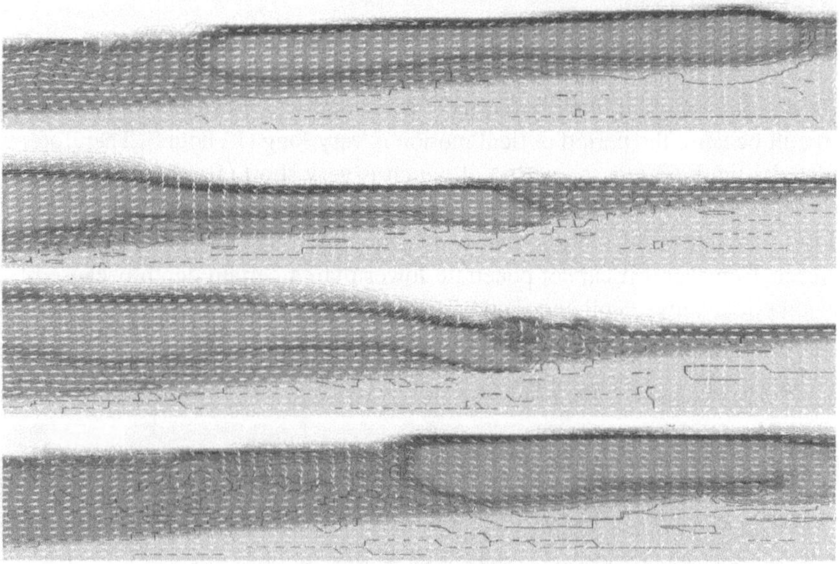

Fig. 5.8. Oil adhesion ($g_T = 0$)

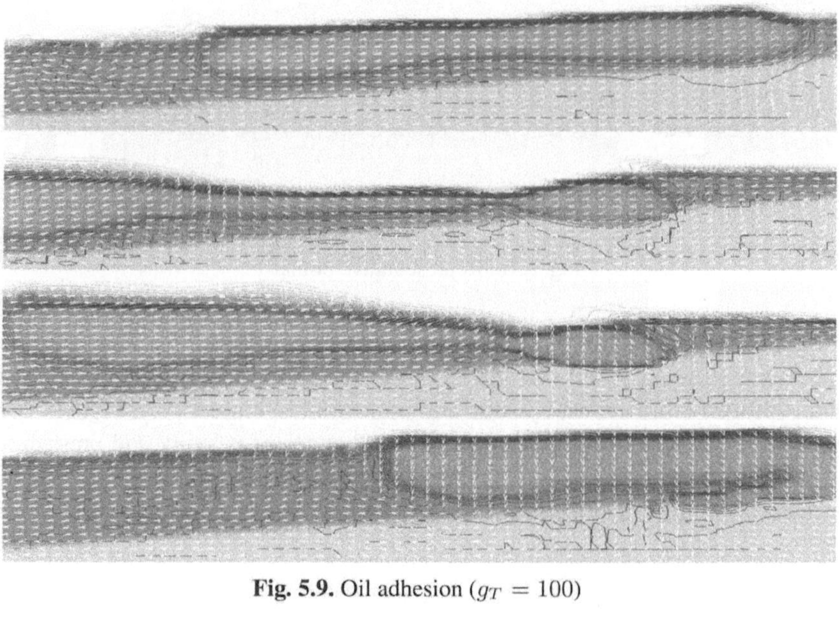

Fig. 5.9. Oil adhesion ($g_T = 100$)

5.5.4 Infiltration phenomena of oil into a tidal flat or sea bed due to the tidal motion

The experimental study stated in Chap. 4 shows that tidal motion plays an important role in penetration of oil into sandy beach. But numerical realization of tidal motion is difficult because the period of tidal motion is very long (12 hours). Therefore, we simulated a tidal motion, the period of which is very short (10 minutes). However, it shows identical tendencies as in the physical experiments.

Fig. 5.11 shows oil motion due to tidal motion. Seawater goes down to low-tide level. However, oil cannot penetrate lower than a certain depth because of an adhesion effect between oil and sand particles. In this state, adhesion force balances gravitational force. See color figures corresponding to these figures included in the appendix.

5.5.5 Blocking effect by penetrated oil on the internal flow in sand

Internal flow in sand is simulated in cases with and without spilled oil. Time variations of velocity and pressure are measured at three points A, B and C in sand, one of which is located near the beach while the other is located deep in the sand. (See Fig. 5.10.)

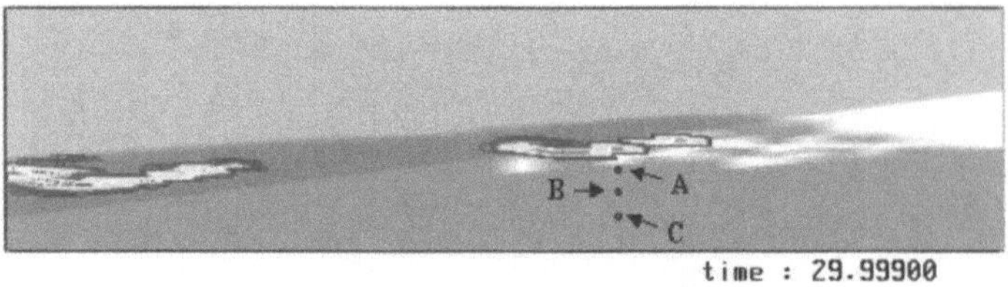

Fig. 5.10. Representative points

In Figs 5.12 to 5.14, time variations of the absolute value of velocity with and without spilled oil are shown by solid and broken lines, respectively. Figures. 5.15 to 5.17 show time variations of the pressure. Changes of velocity and pressure of the internal flow corresponding to above cases are clearly observed in these figures. Influences by spilled oil become smaller as observation point becomes distant from spilled oil.

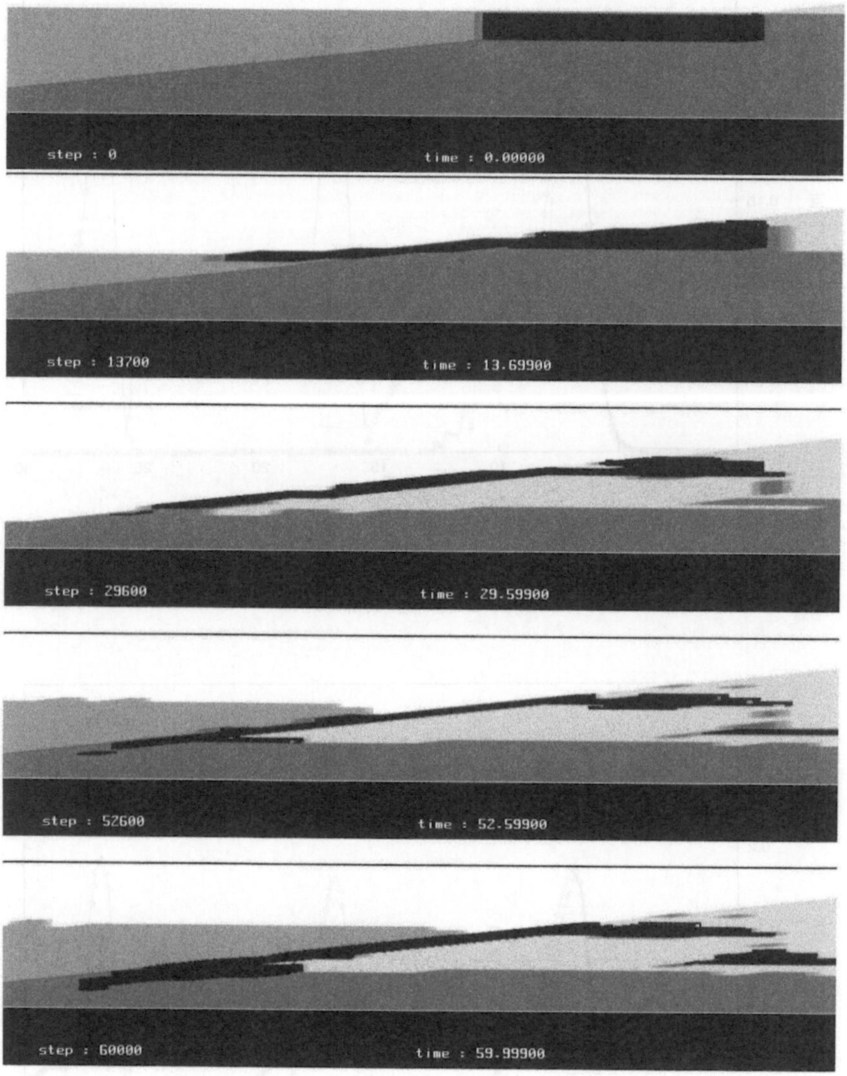

Fig. 5.11. Motion of spilled oil due to tidal motion

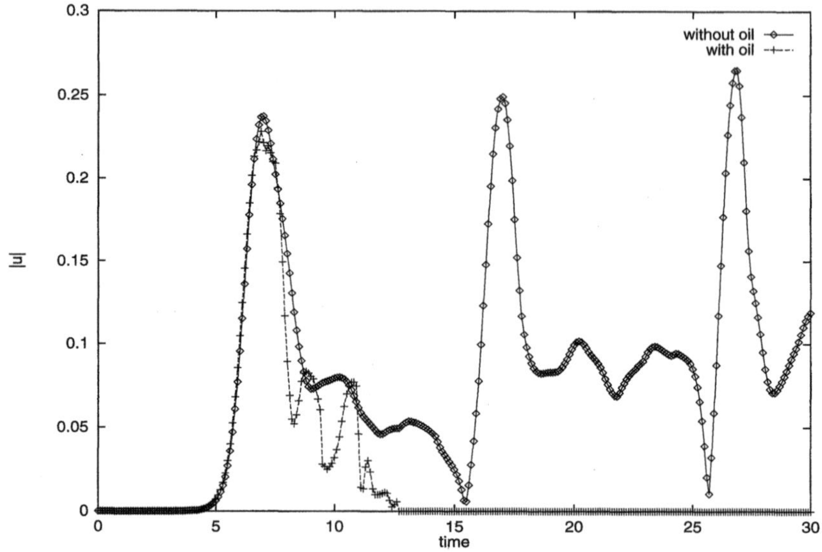

Fig. 5.12. Variation of $|u|$ at point A

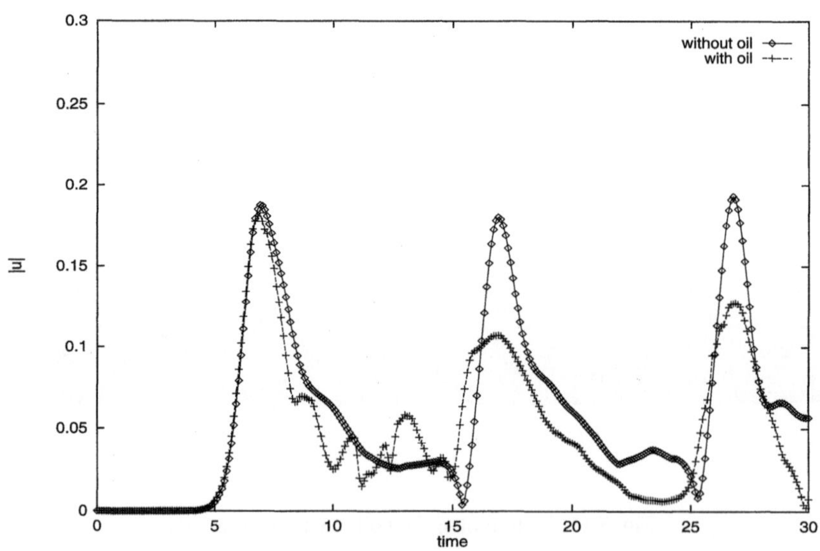

Fig. 5.13. Variation of $|u|$ at point B

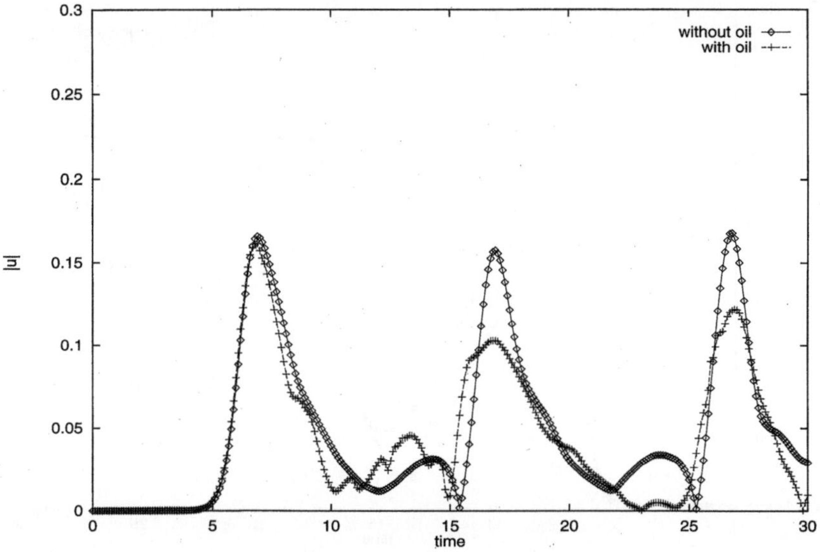

Fig. 5.14. Variation of $|u|$ at point C

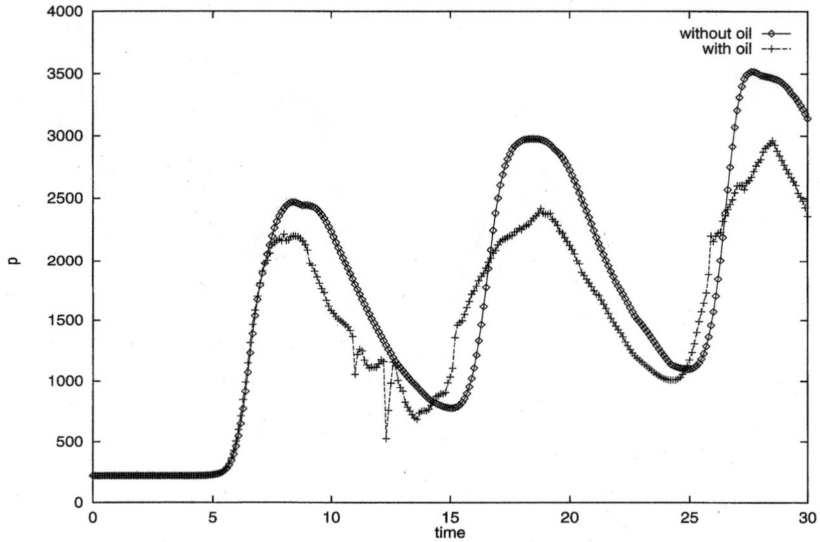

Fig. 5.15. Variation of p at point A

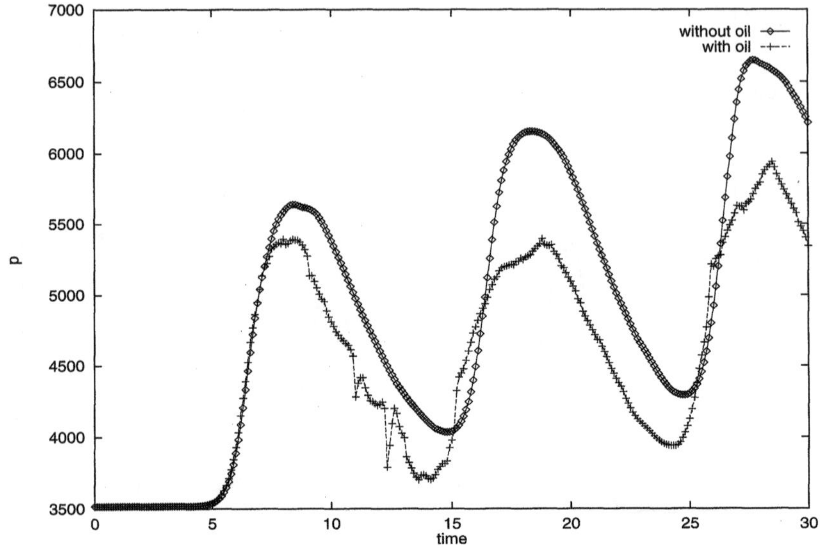

Fig. 5.16. Variation of p at point B

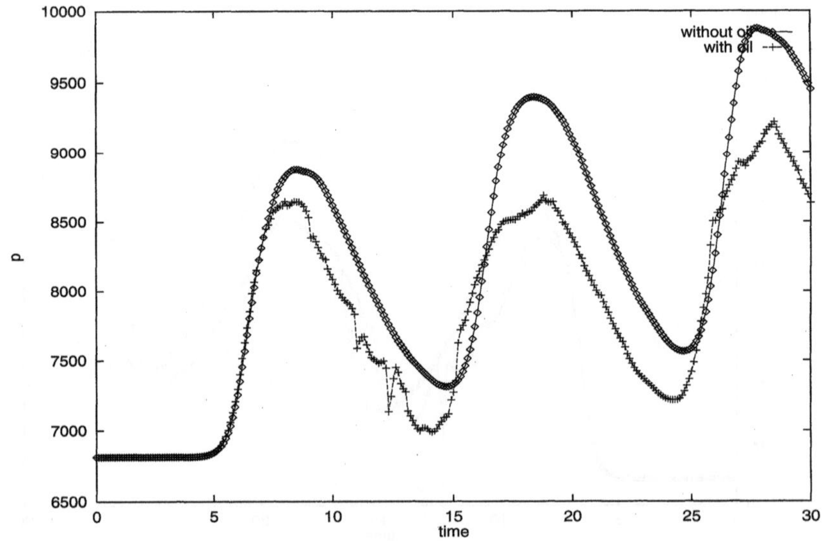

Fig. 5.17. Variation of p at point C

References

Barnes HA, Hutton JF, Walters K (1989) An Introduction to Rheology, Elsevier Science Publishers

Bingham EC (1922) Fluidity and Plasticity, McGraw-Hill, New York

Cross MM (1965) Rheology of non-Newtonian fluids: a new flow equation for pseudo-plastic systems. J Colloid Sci 20:417–437

Duvaut G, Lions JL (1976) Inequalities in Mechanics and Physics, Springer-Verlag

Fujita H, Kawarada H (1998) Variational inequalities for the Stokes equation with the boundary conditions of friction type. GAKUTO International Series Mathematical Sciences and Applications. 11:15–33

Fujita H, Kawahara H, Kawarada H (1995) Distribution theoretic approach to fictitious domain method for Neumann problems. East-West J Numer Math 3:111–126

Glowinski R, Lions JL, and Tremolieres R (1981) Numerical Analysis of Variational Inequalities, North-Holland Publishing Company

Kawarada H, Fujita H, Suito H (1998) Wave motion breaking upon the shore. GAKUTO International Series Mathematical Sciences and Applications 11:145–159

Kawarada H, Suito H (1997) Numerical method for a free surface flow on the basis of the fictitious domain method. East-West J Numer Math 5(1):57–66

Sisko AW (1958) The flow of lubricating greases. Ind Eng Chem 50:1789–1792

6 Oil pollution: human damage on hydraulic regime and benthic communities in tidal flat ecosystems

6.1 Introduction

Waves and tides are two major driving forces to supply seawater into sandy beach and tidal flat sediments. Contrary to the role of waves on seawater infiltration, tides are major force than waves for oil penetration into sediments as shown in Chapter 4. This chapter deals with infiltration of stranded oils into tidal flat sediments and its effects on seawater infiltration.

The same simulator shown in Chapter 2, Fig. 2.1 with wave and tide controls were used packed with the sediment sampled from a natural tidal flat in Fukuyama, Japan (Fig. 6.1). The tidal level was controlled with the range of 40 cm as shown in Fig. 6.2. Tide was controlled with semi-diurnal (12 hrs) tidal cycle and the vertical fluctuation rate of seawater level by tide was 0.009 cm s^{-1} based on the mean tidal range of 2m in the Hiroshima Bay. No tidal fluctuation was given during lag period for out of the tidal range to simulate the natural tidal fluctuation. The sediment contains 2.8 % of silt and clay particles and particle size distribution is shown in Fig. 6.3. The organic content, porosity and mean grain size of the sediment were 0.93%, 19% and 0.65 mm, respectively.

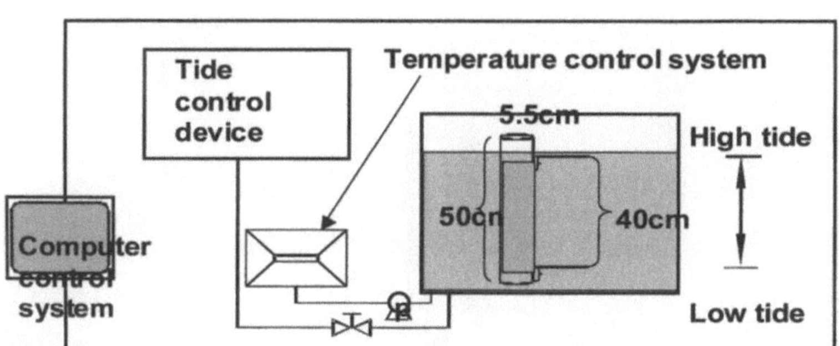

Fig. 6.1. Schematic diagram of experimental set-up for oiling

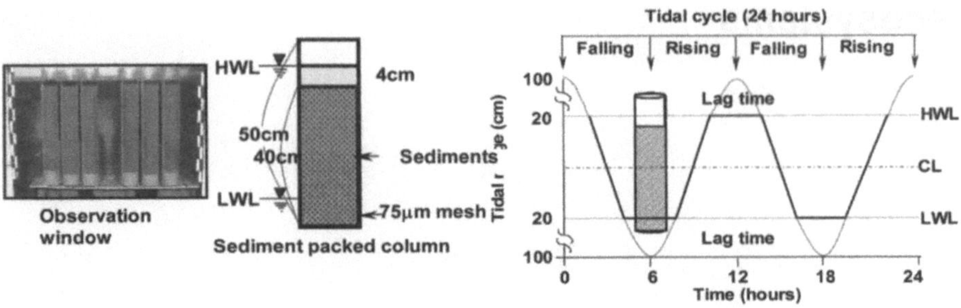

Fig. 6.2. Packed column of tidal flat sediments and the pattern of tidal fluctuation for the columns (HWL; high water level, CL; center line, LWL; low water level)

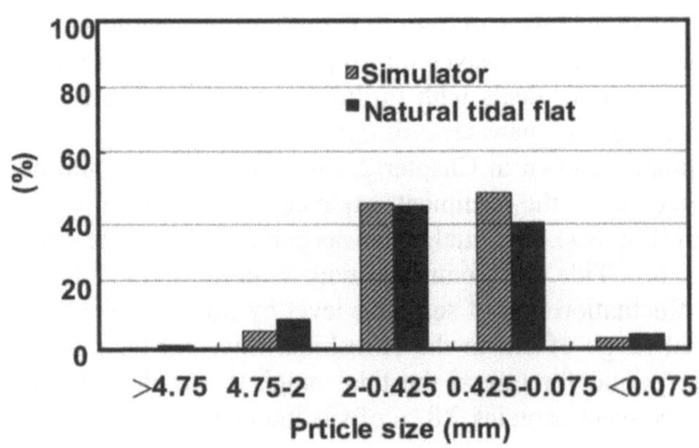

Fig. 6.3. Particle size distribution of the tidal flat sediments

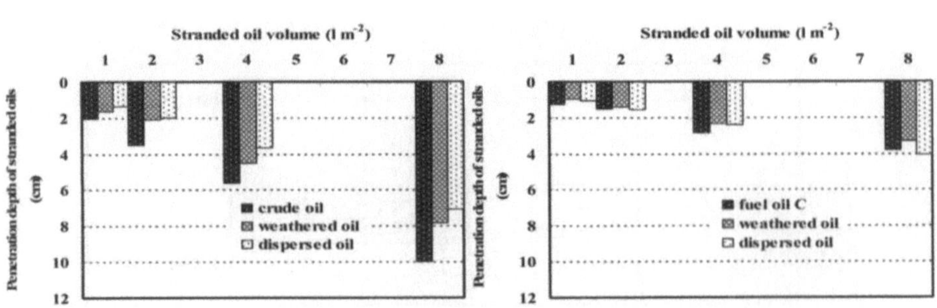

Fig. 6.4. Penetration depths of weathered and dispersed crude oil (left) and fuel oil C (right) after the first tide.

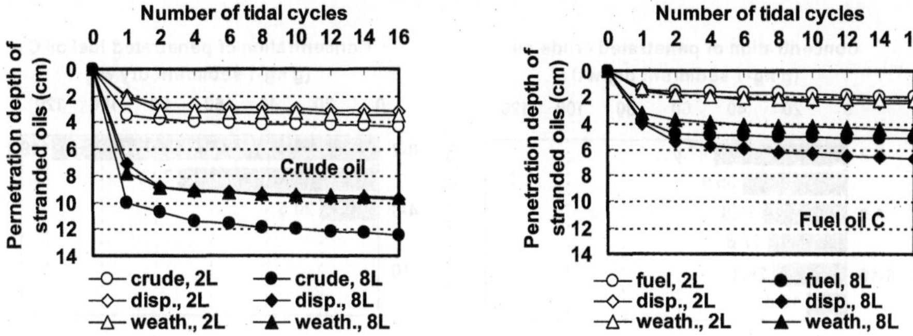

Fig. 6.5. Penetration behavior of oils over 16 tidal cycles

6.2 Penetration of stranded oils into tidal flat sediments by tides

Fig. 6.4 shows penetration depths of fresh, weathered and dispersed oils originated from crude oil and fuel oil C after the first tide from the strand. The depths of penetration into tidal flat sediments were smaller than those into sandy beach. For example, fuel oil C (4 l m^{-2}) penetrated down to 6 cm in sandy beach (see Fig. 4.3), whereas its penetration was less than 3 cm in the tidal flat sediments. The smaller sediment particles and the smaller porosity in the tidal flat sediments may be responsible for the smaller penetration compared to sandy beach sediments. Similar to the case in sandy beach, the depth of penetration was deeper for crude oil than for fuel oil C (See Fig. 4.2). This difference may be due to the difference in viscosity (crude oil : 28 mm^2 s^{-1}, fuel oil C : 3,750 mm^2 s^{-1} at 15℃). Also the increase in the amount of stranded volume increased the depth of penetration (see Fig. 4.3).

Both weathered and dispersed oils did not necessarily show deeper penetration than fresh oils. Especially in the case of crude oil, weathering and dispersion seemed to prevent the penetration oil into the sediments. Penetrations depth of crude oil and fuel oil C over 16 tidal cycles are shown in Fig. 6.5. Similar to the cases in sandy beach sediments (Fig. 4.12), the most significant penetration was noted in the first tidal cycle. Little more penetration in the following tidal cycles was noted after the first cycle. Although the depth increased from 10 cm in the first to 13 cm in the 16th cycle in the case of 8 L m^{-2} of crude oil, little increases were noted for other cases.

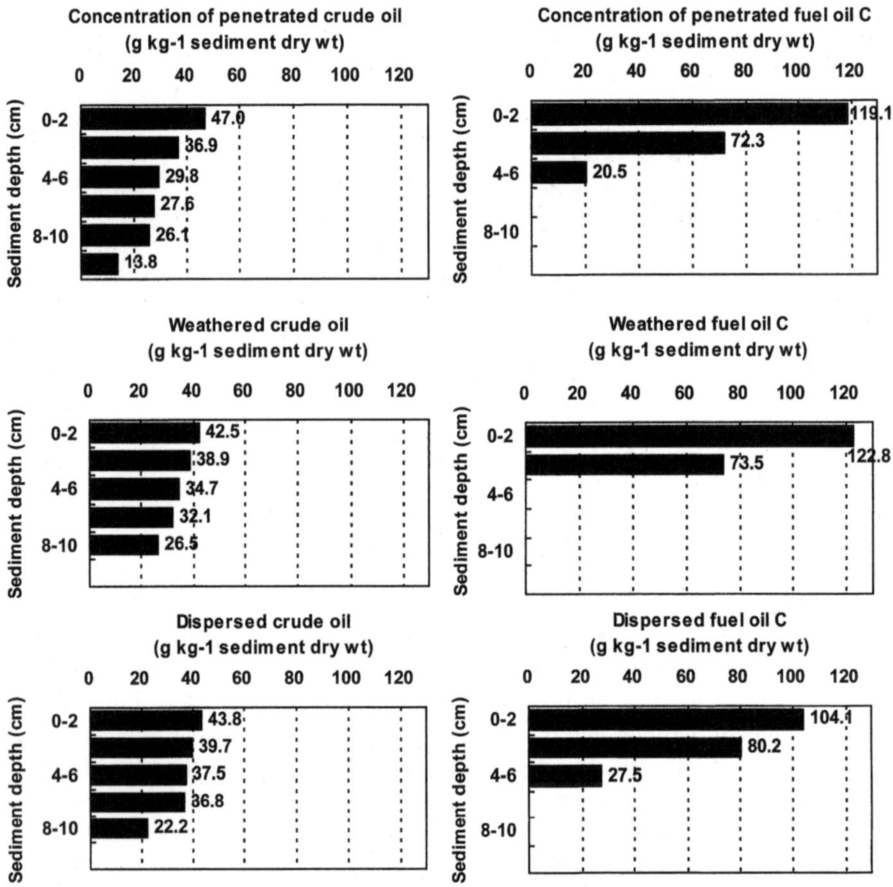

Fig. 6.6. Comparison of vertical distribution of penetrated oils in the condition of applied oil volume as 8L per square meters after 16 tidal cycles

Fig. 6.6 shows the vertical distribution of crude oil and fuel oil C concentration after the 16th tidal cycle when oils were applied for 8 L m^{-2}. The vertical profiles were significantly different between crude oil and fuel oil C irrespective of fresh, weathered and dispersed oils. The surface layer showed the highest concentration both for crude and fuel oils. Although concentrations of fresh, weathered and dispersed crude oils showed gradual decrease along the depth of soils, most of fuel oil C still remained at the surface layers even after 16th tidal cycles. The vertical distributions were significantly different from the case in sandy beach sediments as shown in Fig. 4.13. It is most likely that the fuel oil C with high viscosity was concentrated at the surface layer of tidal flat sediments with fine particles.

6.3 Effects of the penetrated oils into tidal flat sediments on seawater infiltration by tides

Fig. 6.7 shows effects of weathering and dispersant and the amount of stranded oil on the volume of seawater infiltration by tides after the first tidal cycle. The left figure corresponds to crude oil and the right to fuel oil C. The infiltration volume of seawater by tides was 0.061 m^3 m^{-2} day^{-1} without oil pollution. In the first tidal cycle after the strand, it decreased to 0.054 m^3 m^{-2} day^{-1} and 0.045 m^3 m^{-2} day^{-1} for crude and fuel oil C, respectively at the stranded volume of 1 L m^{-2}. As expected, the stranded oil volume showed the most significant effects on the seawater infiltration. The increase in the volume of strand from 1 L m^{-2} to 8 L m^{-2} decreased infiltration volume down to 0.031 and 0.022 m^3 m^{-2} day^{-1} in the case of crude oil and fuel oil C, respectively.

Although, significant differences between crude oil and fuel oil C were not noted, fuel oil C accumulated more in the surface showed larger decrease in infiltration volume than crude oil. Also, the differences among fresh, weathered and dispersed oils were not remarkable in the case of tidal flat sediments. The infiltration volume of dispersed oils was the highest followed by weathered and fresh oils, even though the vertical distribution of oil concentration was almost same shown in Fig.6.6. These are remarkable difference from that in sandy beach sediments (Figs. 4.16, 4.20), i.e. the more significant differences were noted between crude oil and fuel oil C and among fresh, weathered and dispersed oils.

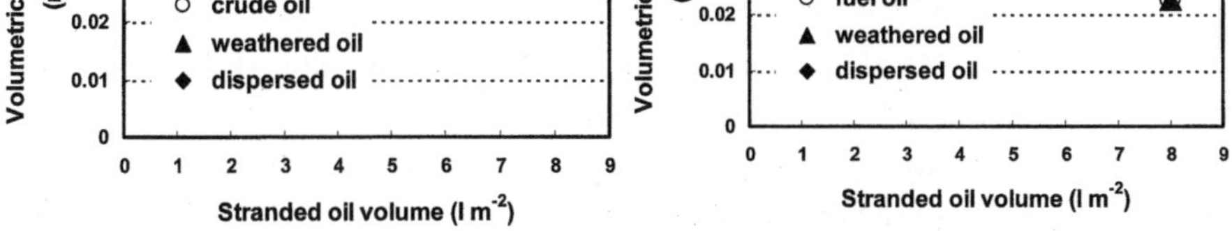

Fig. 6.7. Seawater infiltration through the oiled tidal flat sediments (left: fresh, weathered and dispersed crude oil; right: fresh, weathered and dispersed fuel oil C

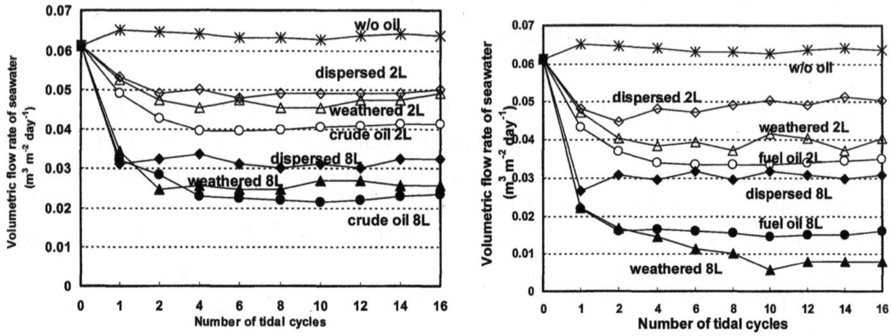

Fig. 6.8. Seawater infiltration through the oiled tidal flat over 16 tidal cycles (left: fresh, weathered and dispersed crude oil; right: fresh, weathered and dispersed fuel oil C)

Fig. 6.8 shows effects of stranded fresh, weathered and dispersed crude oil and fuel oil C on seawater infiltration by tides as affected by tidal cycles. The largest decreases in the infiltration volume were noted during the first tidal cycle. Little differences were noted after the third cycle until the 16th cycle. It may take longer time to recover the original rate of infiltration by washout and biodegradation of these oils.

Fig. 6.9 shows the relationship between seawater infiltration rate and clogging of pore volume in the surface sediments (0 cm to 2 cm) by oil. The plots and the line indicate the experimental observations shown in Fig. 6.7 and estimated rates of seawater infiltration with a modified Kozeny's equation as follows (Stevenson, 1997);

$$u = \frac{\tau}{k_0} \cdot \frac{\varepsilon^3}{\left(1-\varepsilon^2\right)} \cdot \frac{1}{\mu S_v^2} \cdot \frac{\Delta p}{L} \tag{6.1}$$

where, u is the rate of infiltration [cm s^{-1}]; τ is the turtosity [-]; k_0 is a constant [-]; ε is the porosity [-]; μ is the absolute viscosity [cm^2 s^{-1}]; S_v is the specific area of sediment particles [cm^2 cm^{-3}]; Δp is the pressure loss [(g cm s^{-2}) cm^{-2}]; and L is the length of the column [cm].

Eq. (6.1) was modified for the seawater infiltration into the sediments clogged with oil. The porosity, ε, is dependent on the percent of the pore volume clogged by oil, ξ(%), and is replaced by $(1- \xi/100)\,\varepsilon$. The viscosity of seawater, pressure loss and column length were kept constant in the experiment. Therefore $(1/k_0)\cdot(1/\mu)\cdot \Delta p/L$ was replaced by a constant a (i.e. $a = (1/k_0)\cdot(1/\mu)\cdot \Delta p/L$). Thus, Eq. (6.1) is modified as follows,

$$u = a \cdot \tau \cdot \frac{1}{S_v^2} \frac{(\varepsilon - \xi / 100\varepsilon)^3}{\left(1 - (\varepsilon - \xi / 100\varepsilon)^2\right)} \tag{6.2}$$

For the unoiled case, $u = 0.0151$ [cm s^{-1}], ε=0.43 [-], and ξ=0 [-],

$$a \cdot \tau \cdot \frac{1}{S_v^2} = 0.155 \tag{6.3}$$

Therefore, Eq. (6.2) can be expressed as follows assuming that there are no changes in the turtosity and the specific surface area of the sediment particles with the increase in ξ.

$$u = 0.155 \frac{(\varepsilon - \xi/100\varepsilon)^3}{(1 - (\varepsilon - \xi/100\varepsilon)^2)} \tag{6.4}$$

The relationship between u and ξ based on Eq. (6.4) is also shown in Fig. 6.9. It is clear that the clogging of pores is the major factor to decrease seawater infiltration rate rather than oil constituents like crud oil or fuel oil C. The figure shows that if a plot is located in the upper or lower part of Eq. (6.4), the apparent turtosity or the specific surface area has been changed. The plots for fuel oil C are located in the upper part of the curve indicating that the clogging by fuel oil C increased apparent turtosity or decreased specific surface area.

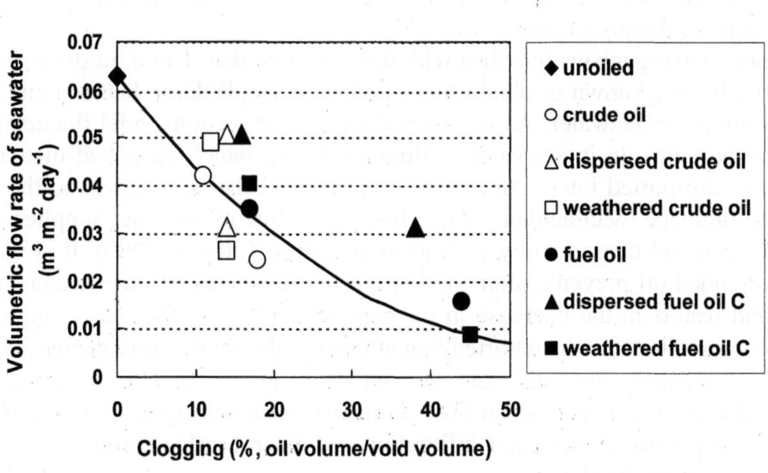

Fig. 6.9. Effects of clogging (% in volume) of void in the sediments on the rate of seawater infiltration by tides

6.4 Effects of oil spill on seawater infiltration and macrobenthic community in tidal flats

6.4.1 Introduction

Many oil spill accidents by tankers for transportation of petroleum have occurred all over the world. A part of the spilled oils into sea is drifting toward coastal area. Approximately a half of the spilled oil was transferred by winds to the coast in the Nakhodka oil spill in Japan (Hozumi, 2000). Wolfes (1994) also reported that about a half of 42 million liters of the oil spilled from the Exxon Valdez was stranded on the shoreline of the Prince William Sound.

In spilled oil stranded on tidal flat, which are important habitats for many commercial fishes and migratory birds, the oil pollution of tidal flats may result in tremendous damages on tidal flat ecosystem. Toxicological risk of pollutants on ecosystems is evaluated by their fate (its transfer and transformation in the environment) and effects (its impact on biological communities) in the ecosystem. Studies on biological effects include identifying and quantifying instances of exposure of biological communities to spilled hydrocarbons, evaluating acute, chronic and residual toxicity to organisms, and assessment of subsistent food contamination. The primary mechanisms by which petroleum hydrocarbons are deleterious to benthic organisms include oxygen stress (from organic enrichment) and direct toxic or carcinogenic effect on organisms (Connell & Miller, 1984). The ecological damages in tidal flats may be caused not only by the toxicity of oil constituents but also by the changes in physico-chemical condition like infiltration of seawater into tidal flat sediments.

Macrobenthic organisms (benthos with the size more than 1 mm mesh size) living in tidal flat are known to obtain their food, such as plankton, bacteria and detritus, mainly from seawater. As shown in the previous sections, tidal fluctuations play an important role in seawater infiltration during which dissolved matter in seawater is transported into benthic ecosystem in tidal flat as it returns to the sea by gravity drainage (McLachlan, 1982). Inter tidal flow of seawater supplies dissolved oxygen and consequently biological activities (Cheong, 2000). It is likely that the stranded oil prevents interstitial spaces of sediments, reduces water infiltration, and results in the decrease in oxygen, nutrients and other food supply to benthic communities. It was reported that stranded oils significantly decreased infiltration of seawater into tidal flat sediment (Cheong, 2000). This section deals with the effects of stranded oil on tidal flat ecosystems with special emphasis on the relationship between seawater infiltration and macrobenthic community using tidal flat ecosystem simulators.

6.4.2 The tidal flat simulator

Fig. 6.10 is a picture of the tidal flat simulator (Japan Aquatic Tech. Co. Ltd., JBES-2500W) used in this study. The simulator is similar to the one in Fig. 2.1. The tidal flat in the simulator is 4.5 m long and 0.8 m wide with a slope of 3/100. Two simulators with the same configuration were used for this study. These simulators are designed as a typical tidal flat ecosystem in the Seto Island Sea, Japan. The tidal flats were operated with a tidal fluctuation of 0.009 cm s^{-1}, wave height of 30 mm and seawater temperature at 20 ℃. Tidal velocity of 0.009 cm s^{-1} was determined from the mean tidal range of 2 m in Hiroshima Bay. Breaking wave height and wave periods in this study were set to 30mm and 0.8 sec, respectively (Hayes, 1999). The sediment in the tidal flat simulator was collected from a natural tidal flat in Hiroshima Bay, Japan. The contents of silt & clay (< 0.075 mm) in the sediments of C and O simulators were 2.9% and 3.7%, respectively. Those in natural tidal flats in Hiroshima Bay, Seto Inland Sea, ranged from 2.7% to 8.7% Japan (Lee, 1998). Macrobenthos populations were incubated for more than one month from an inoculum collected in the natural tidal flat. Fuel oil C (1 L m^{-2}) was spilled onto a tidal flat simulator (O simulator). Another simulator was operated as a control without any oil spill (C simulator). Vertical profiles of oxidation-reduction potential (ORP) in the tidal flat sediments were continuously monitored with intervals of 5 or 1cm by a multi-point electrode (TOA electronics Ltd., HM-50G) and a reference electrode (TOA electronics Ltd., HM205C) with saturated aqueous solution of silver chloride.

Fig. 6.10. Tidal flat ecosystem simulators (O and C simulators) (a) Tidal flat, (b) Wave maker, (c) Tide control device, (d) Seawater tank, (e) Temperature control system, (f) Computer control system

Seawater and sediments were prepared to have the same conditions as those in natural tidal flats. Seawater was sampled to determine dissolved inorganic nutrients ((NO_2+NO_3)-N, PO_4-P) and chlorophyll-*a* concentration. Sediments were also sampled to determine chlorophyll-*a* and macrobenthic populations. Volume of seawater infiltration by tide was estimated from the difference in water content of the sediments between high and low tides according to the method described by Cheong (2000). Concentration of hopane, non-biodegradable constituents in the oil, in the sediment was determined to estimate washout and biodegradation of the spilled oil. Total number of bacteria in sediment was counted according to the direct counting method with DAPI (4, 6-diamidino-2-phenylindole) (Porter and Feig, 1980).

The simulators were operated for 30 days prior to the oil spill. Fuel oil C was spilled into O simulator at day 0, whereas no spill was occurred in C simulator. Figure 6.11 shows dissolved inorganic nitrogen (a) and phosphorus (b) concentrations in seawater before and after the oil spill in the O and C simulators. (NO_2+NO_3)-N and PO_4-P concentrations ranged from 0.1 to 1.1 μg-at N l^{-1} and from 0.02 to 0.34 μg-at P l^{-1}, respectively. These concentrations are similar to those in Hiroshima Bay, i.e. from 0.2 to 10 (μg-at N l^{-1}), and from 0.1 to 1.5 (μg-at P l^{-1}), respectively, indicating that the water quality in the simulator is similar to that in Hiroshima Bay.

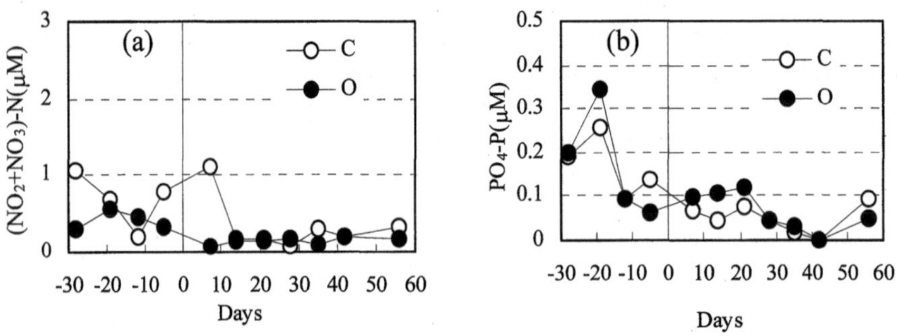

Fig. 6.11. Variation of (NO_2+NO_3)-N (a) and PO_4-P (b) concentrations in seawater

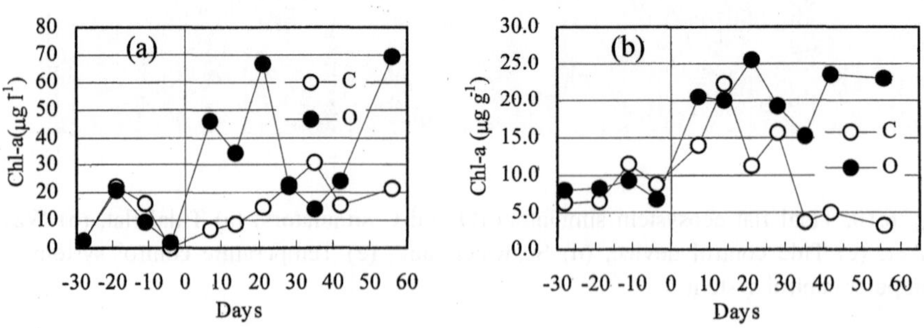

Fig. 6.12. Chlorophyll-*a* concentrations in seawater (a) and sediments (b) in the C and O simulators

Fig. 6.12 shows chlorophyll-*a* concentrations both in seawater and surface sediment in the C and O simulators. Chlorophyll-*a* concentrations in seawater and sediments in the C simulator ranged from 0.29 to 30.8 μg l^{-1}, and from 3.2 to 22.2 μg g^{-1}, respectively. These are almost same levels as those in Hiroshima Bay (Lee, 1996) indicating primary productivity in the simulators were enough to support benthic populations.

Fig. 6.13 shows vertical profiles of ORP in the C and O simulators. ORP in the surface sediments (1cm) were positive (average: 250 mV) throughout the experimental period, whereas those in 5cm deep were negative (average: - 400 mV). These values were similar to these in natural tidal flats in Hiroshima Bay (average: + 134 mV in 2cm deep and –60 mV 5cm deep) (Lee, 1996). After the oil spill, ORP in the O simulator dropped down to negative values even in the surface layer. The accumulation of oil in the surface sediments seemed to prevent oxygen supply from seawater and air into sediments and/or enhanced oxygen consumption by bacteria during oil decomposition in the sediments.

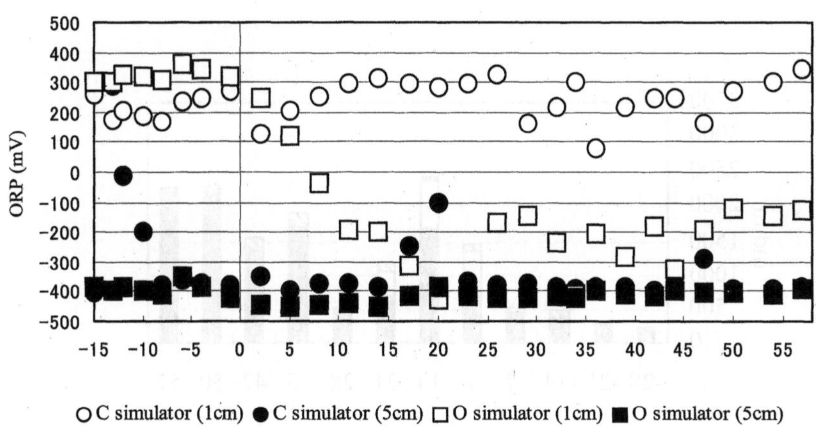

Fig. 6.13. Effect of oil spill on ORP in C and O simulators

Fig. 6.14 shows bacterial population in the sediments of C and O simulators. The bacterial population in the C simulator and O simulator ranged from 0.6 to 2.5 x 10^9 cells g^{-1} and were similar to those in the sediment in Hiroshima Bay, i.e. 1.6 to 3.6 x 10^9 cells g^{-1}. Significant changes in bacterial density were not noted after the oil spill.

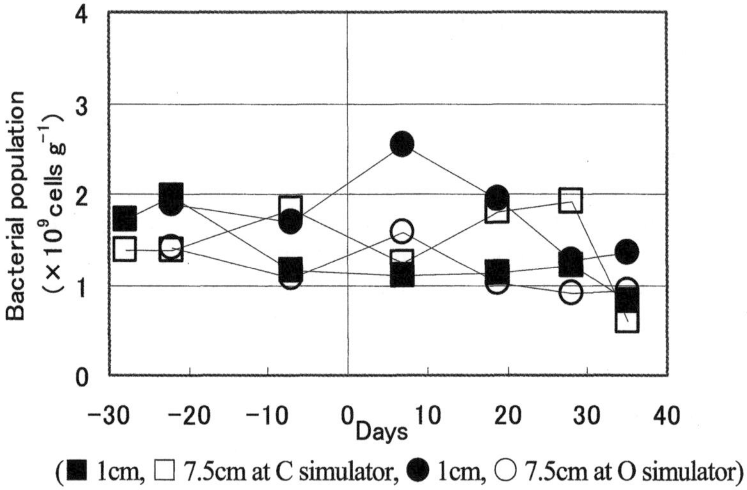

Fig. 6.14. Bacterial population in C and O simulators

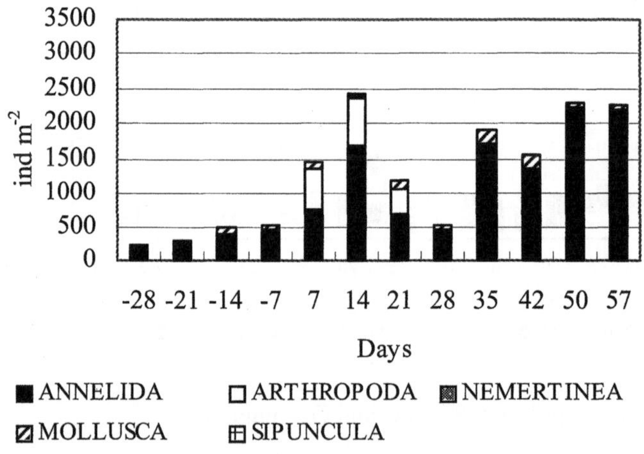

Fig. 6.15. Total population density of macrobenthos in the C simulator

Figure 6.15 shows population density of macrobenthos in the C simulator. In the C simulator before day 0, total population density of macrobenthos ranged from 225 to 525 m^{-2}, and the dominant species were *Capitella* sp., *Ceratonereis erthraeensis* and *Mediomastus* sp. After a month, total population density of macrobenthos increased from 1200 to 2400 m^{-2} and was kept high throughout the experimental period indicating a formation of a stable ecosystem. Total population densities of macrobenthos in natural tidal flats, Hiroshima Bay, were reported to range from 1760 to 6784 m^{-2} and dominant species were Polychaeta (Lee, 1998).

As shown above, DIN and DIP concentrations in seawater and chlorophyll-*a* concentrations in seawater and sediment in the C simulator were similar to those in the coast area in Hiroshima Bay. ORP and Bacterial population were also similar to those in the Hiroshima Bay. Although population densities of macrobenthos in the C simulator were lower than those in Hiroshima Bay, these results suggest that the ecosystem maintained the C simulator represents the tidal flat ecosystems typical in Hiroshima Bay area.

6.4.3 Effects of oil spill on tidal flat ecosystem

The penetrated oil concentrations in the sediments after the spill were estimated from the concentration of hopane, i.e. a non-degradable fraction in oil and usually being used as a biomarker. Figure 6.16 shows vertical profiles of the oil concentrations in the sediments in the O simulator. The oil concentration in the surface sediments decreased from 0.9 in day 7 down to 0.5 l m^{-2} in 35 days, whereas they were less than 0.01 lm^{-2} in deeper layers. The reduction may be due to biodegradation and washout bay tides. However, significant increase in bacterial population in the surface sediment was not noted throughout the experimental period as shown in Fig 6.14. It is most likely that the oil decrease is due to the washout, i.e., release of the oil into seawater by tidal and wave actions rather than biodegradation.

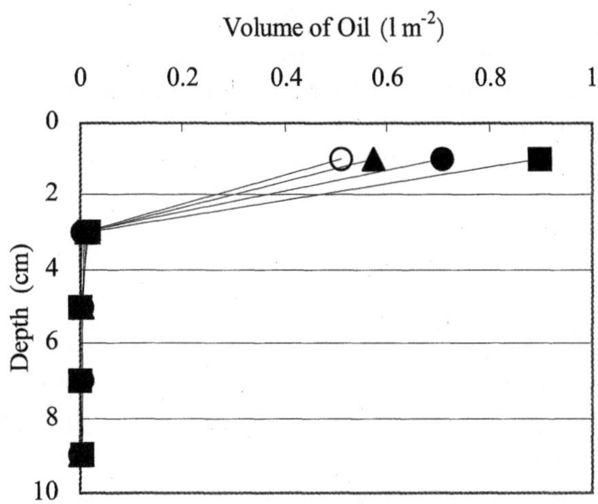

Fig. 6.16. Vertical profiles of fuel oil C concentration in O simulator sediments (■ 7 days, ● 14 days, ▲ 28 days, ○ 35 days after the oil spill)

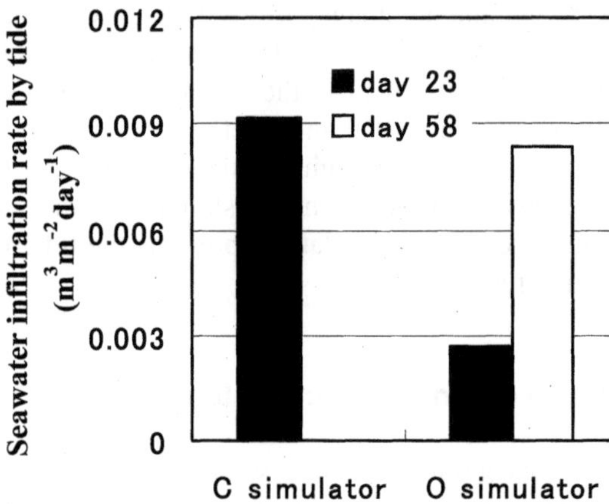

Fig. 6.17. Infiltration rate of seawater by tide at day 23 and 58 in the C and O simulators

Figure 6.17 shows infiltration rate of seawater by tidal action both in the C and O simulators. The infiltration rates of sweater at day 23 in the C and O simulators were 9.2×10^{-3} and 2.7×10^{-3} m^3 m^{-2} day^{-1}, respectively. The oil spill and the penetration into the sediments decreased infiltration rate of seawater down to a third. The accumulation of fuel oil C on the tidal flat sediment (Fig.6.16) is mainly responsible for the decrease in the filtration rate. Also, the growth of benthic algae (Fig.6.12(b)) seems to be responsible for the reduction of infiltration rate. The concentration of chlorophyll-*a* in the surface sediment in day 23 was higher than that before the oil spill. In day 58, the infiltration rate in the O simulator returned to that without oil spill, i.e. 8.5×10^{-3} m^3 m^{-2} day^{-1}. The restoration of the infiltration rate is consistent with that of macrobenthic population density as shown later on. The increase in the infiltration rate is probably due to the decrease in the fuel oil C on the surface sediment.

6.4.4 Macrobenthic community

Fig. 6.18 shows population density of macrobenthos in the O simulator. In the O simulator, the population density was kept low after the oil spill until day 35, whereas the higher population density was observed in the C simulator was kept higher (Fig. 6.15). These results suggest that the oil spill gave a significant impact on the macrobenthic population. David (1996) reported that physical factors rather than chemical toxicity of the spilled oil affected inter-tidal fauna. Major reason for the reduction of benthic population here in this study also seemed to be physical factor; the oil accumulated in the sediment surface resulted in the decrease in ORP and infiltration rate of seawater. In day 42 from the oil spill, the macrobenthic community was almost recovered. This recovery is coincident with the recovery of the infiltration rate of seawater as shown in Fig. 6.17. These results indicate that the infiltration of seawater with dissolved oxygen, nutrients and organic matters plays an important role in the survival of macrobenthic community.

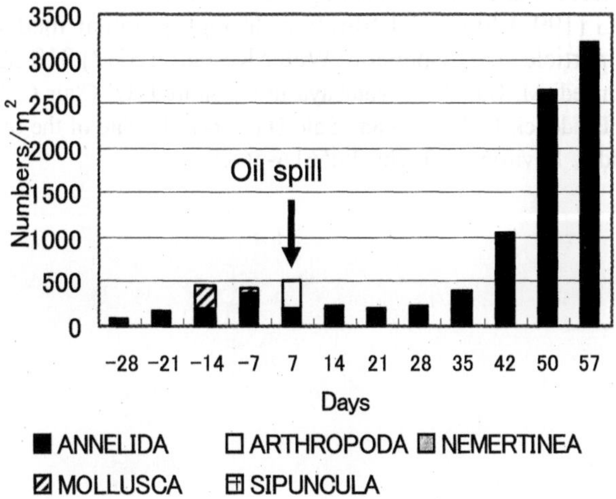

Fig. 6.18. Population of macrobenthos in the simulators through the experimental period

References

Cheong CJ, Lee JG, Nishima W, Baba E, and Okada M (2000) Seawater infiltration into tidal flat or sandy beach sediment by wave action in enclosed bay. J Japan Society on Water Environ 23(10):619-623

Hayes MO and Michel J (1999) Factors determining the long-term persistence of Exxon Valdez oil in gravel beaches. Marine Pollution bulletin 38:92-101

Hozumi T, Tsutsumi H and Kono M (2000) Bioremediation on the shore after an oil spill from the Nakhodka in the sea of Japan. Marine Pollution. Bulletin 40(4):308-314

Lee JG, Nishijima W, Mukai T, Takimoto K, Seiki T, Hiraoka K and Okada M (1997) Comparison for structure and functions of organic matter degradation at natural and constructed tidal flats - A case study in Hiroshima bay -. J Japan Society on Water Environ 20(3):175-184

Lee JG, Nishijima W, Mukai T, Takimoto K, Seiki T, Hiraoka K and Okada M (1998) Factors to determine the functions and structures in natural and constructed tidal flats. Wat Res 32(9):2601-2606

Lee YS (1996) The mechanism of phytoplankton growth in Hiroshima Bay. Ph.D. Thesis, Hiroshima Univ.

McLachlan A (1982) A Model for the Estimation of Water Filtration and Nutrient Regeneration by Exposed Sandy Beaches. Marine Environmental Research 6: 37-47

Porter KG and Feig YS (1980) the use of DAPI for identifying and counting aquatic microflora. Limnol Oceanogr 25:119-125.

Stevenson, DG (1997) Flow and filtration through granular media - the effect of grain and particle size dispersion. Water Research 31(2):310-322

Wolfe DA, Hameedi M, Galt JJA, Watabayashi G, Short J, O' Clair C, Rice S, Michel J, Payne JR, Braddock J, Hanna S and Sale D (1994) The fate of the oil spilled from the Exxon Valdez. Environ Sci Techn 28:561A-568A

7 Decomposition mechanism of spilled oil by bacteria

7.1 Introduction

The unified model formulations developed in Chapters 3 and 5 make it possible to clarify mechanisms of oil decomposition by adding crucial devices to them; this has not been experimentally clarified yet. Multi-phase flow formulation with decomposition contributes to this study. Main components of decomposed materials by bacteria are water, soluble components, and carbon oxides. We deal with the case in which oil decomposes into water. A characteristic of this case is appearance of a jump boundary condition for normal component of velocity at the decomposition surface between oil and water, which is based on density difference between them.

7.2 Notations

x_i	: Cartesian coordinates $(i = 1, 2, 3)$
u_i	: Velocity vector $(i = 1, 2, 3)$
u_n	: Normal component of velocity on boundary
u_T	: Tangential component of velocity on boundary
p	: Pressure
ρ	: Density
t	: Time
D_{ij}	: Rate of strain tensor
Ω_α	: Domain for α-phase, where $\alpha = \{air,\ water,\ oil\}$
$\Omega = \bigcup_\alpha \overline{\Omega_\alpha}$	: Total domain
Ω_L	: Domain occupied with sand
Ω_U	: $\Omega \backslash \overline{\Omega_L}$
Ω_b	: Habitat of bacteria,
χ_α	: Characteristic function representing a domain Ω_α, i.e., $\chi_\alpha = 1$ in $\overline{\Omega_\alpha}$ and $\chi_\alpha = 0$ in $\Omega \backslash \overline{\Omega_\alpha}$
χ_L	: Characteristic function representing a domain Ω_L
χ_U	: Characteristic function representing a domain Ω_U
χ_b	: Characteristic function representing a domain Ω_b

χ_α^L	: $\chi_L \cdot \chi_\alpha$
χ_α^U	: $\chi_U \cdot \chi_\alpha$
ν^α	: Kinematic viscosity of α-phase
μ^α	: Viscosity of α-phase
σ_{ij}^α	: Stress tensor of α-phase
ρ^α	: Density of α-phase
c	: Coefficient of resistance force received from sand
g_T	: Friction coefficient between oil and water in sand
K_i	: ith component of external force
Γ_I	: Inflow boundary
Γ_S	: Slip boundary
Γ_N	: No-slip boundary
Γ_A	: Outflow boundary
Γ_{OS}	: Boundary of oil in sand, i.e., $\Gamma_{OS} = \partial\Omega_{oil} \cup \Omega_L$
n_i^α	: Outward normal vector defined on $\partial\Omega_\alpha$
T_i^α	: Tangential vector defined on $\partial\Omega_\alpha$
$\mathcal{D}(\Omega)$	: Set of infinitely differentiable functions with support included in Ω.

7.3 Characteristics of oil decomposition into water

Assume that $\Omega_{oil} \cap \Omega_b \neq \{0\}$, which represents an occurrence of decomposition. The interface motion equation for Γ_{OS} between Ω_{oil} and Ω_{ws} is described as

$$\frac{\partial \chi_{oil}}{\partial t} + (u_j^{oil} + \chi_b \, k_{oil} \, n_j^{oil}) \frac{\partial \chi_{oil}}{\partial x_j} = 0 \text{ in } \Omega,\ t > 0, \tag{7.1}$$

or

$$\frac{\partial \chi_{water}}{\partial t} + (u_j^{water} + \chi_b \, k_{water} \, n_j^{water}) \frac{\partial \chi_{water}}{\partial x_j} = 0 \text{ in } \Omega,\ t > 0. \tag{7.2}$$

Here, k_{oil} and k_{water} are rates of consumption for oil and production for water, respectively. Therefore, k_{oil} is a negative constant and k_{water} is a positive constant; χ_b is a characteristic function of Ω_b.

Remark

It holds that

(1) $\rho_{water} > \rho_{oil}$, *and*
(2) $\rho_{water} \cdot k_{water} + \rho_{oil} \cdot k_{oil} = 0$ *(Mass transference condition).*

Then, 1 and 2 follow $-k_{oil} > k_{water} > 0$.

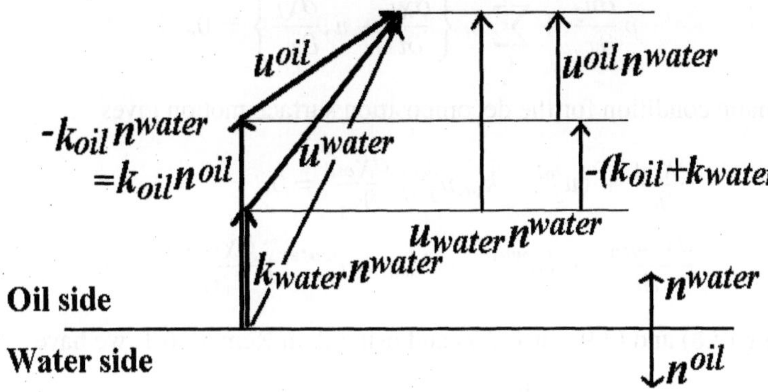

Fig. 7.1. Reactivity condition

7.3.1 Reactivity condition

We assume a continuity of resultant velocity on the boundary between oil and water, i.e.,

$$u_j^{oil} + \chi_b \, k_{oil} \, n_j^{oil} = u_j^{water} + \chi_b \, k_{water} \, n_j^{water} \text{ on } \Gamma_{OS}, \; (j = 1, 2, 3), \quad (7.3)$$

from which follows,

$$\begin{aligned} \left[u_j n_j^{water}\right] &= u_j^{water} n_j^{water} - u_j^{oil} n_j^{water} \\ &= -(k_{oil} + k_{water})\chi_b \text{ on } \Gamma_{OS}, \end{aligned} \quad (7.4)$$

and

$$\left[u_j T_j^{oil}\right] = u_j^{oil} T_j^{oil} - u_j^{water} T_j^{oil} = 0 \text{ on } \Gamma_{OS}. \quad (7.5)$$

Equations (7.4) and (7.5) are called the reactivity condition and $-(k_{oil} + k_{water})$ is the reactivity constant. The meaning of the reactivity condition may be understood through Fig. 7.1.

7.4 Incompressibility condition for a total flow system

Mass conservation law leads to

$$\frac{\partial \rho}{\partial t} + \frac{\partial}{\partial x_j}(\rho u_j) = 0 \text{ in } \Omega, \; t > 0. \quad (7.6)$$

By using $\rho = \sum_l \rho_l \chi_l$, from (7.6) we have

$$\rho \frac{\partial u_j}{\partial x_j} + \sum_l \rho_l \left\{ \frac{\partial \chi_l}{\partial t} + u_j \frac{\partial \chi_l}{\partial x_j} \right\} = 0. \tag{7.7}$$

The kinematic condition for the decomposition surface motion gives

$$\frac{\partial \chi_{oil}}{\partial t} + \left(u_j^{oil} + k_{oil} n_j^{oil}\right) \frac{\partial \chi_{oil}}{\partial x_j} = 0, \tag{7.8}$$

$$\frac{\partial \chi_{water}}{\partial t} + \left(u_j^{water} - k_{water} n_j^{water}\right) \frac{\partial \chi_{water}}{\partial x_j} = 0. \tag{7.9}$$

Substituting (7.8) and (7.9) into (7.7) and using 2. in Remark 6.1, we have

$$\frac{\partial u_j}{\partial x_j} = 0 \ \text{ in } \Omega, \ t > 0, \tag{7.10}$$

which means

$$\int_\Omega \frac{\partial u_j}{\partial x_j} \varphi dx = 0 \text{ for any } \varphi \in \mathcal{D}(\Omega). \tag{7.11}$$

From (7.11), it follows that;

$$\int_{\Omega_o} \frac{\partial u_j}{\partial x_j} \cdot \varphi \, dx + \int_{\Omega_w} \frac{\partial u_j}{\partial x_j} \cdot \varphi \, dx + (k_{oil} + k_{water}) \int_\Gamma \varphi \, d\Gamma = 0, \tag{7.12}$$

which means

$$\frac{\partial \widetilde{u_j}}{\partial x_j} + (k_{oil} + k_{water})\, \delta_\Gamma = 0 \ \text{ in } \Omega. \tag{7.13}$$

Equation (7.13) is the incompressibility condition for a total flow system with decomposition. ($\widetilde{u_j}$ is a restriction of u_j on Ω_{oil} and Ω_{water})

7.5 Biological contribution to satisfy reactivity condition

We propose the following conjecture as one possible interpretation of adding dipole moment distribution along the decomposition surface. Let us consider that many bacteria adhere to the surface of oil; some of them eat oil and the others do not eat it. They contact each other and make pair sets. These pairs are located along the surface of oil. Then there exists differential buoyancy between the pair. Torque caused by this difference produces the distribution of dipole moment on the decomposition surface.

Let γ be a reaction surface which is assumed to be smooth. Then, the dipole moment defined on γ is represented by $\frac{\partial}{\partial \nu}\delta(\gamma)$, where $\delta(\gamma)$ is a delta measure supported on γ. On the other hand, we have $\frac{\partial}{\partial \nu}\delta(\gamma) = \Delta\chi$. Define the volume of a bacteria by V_B. Then we can conclude that

$$(\rho_{water} - \rho_{oil}) V_B \cdot g = -(k_{oil} + k_{water}) \cdot \frac{\rho_{oil}}{k_{water}} V_B \cdot g. \tag{7.14}$$

7.6 Unified model for three-phase flow with decomposition of oil into water

The incompressibility condition for multi-phase flow system is defined as

$$\frac{\partial u_j}{\partial x_j} + (k_{oil} + k_{water})\chi_b \frac{\partial \chi_{oil}}{\partial n_{oil}} = 0, \quad \text{in}\, \Omega, \quad t > 0. \tag{7.15}$$

Existence of the second term in the left hand side is due to the reactivity condition.

Conservation of momentum for the total flow system is established as a unified equation in the following way (See section 5.3.3):

$$\begin{aligned}
\rho\left(\frac{\partial u_i}{\partial t} + u_j \frac{\partial u_i}{\partial x_j}\right) + c\chi_L u_i = &-\frac{\partial p}{\partial x_i} \\
&+\frac{\partial}{\partial x_j}\left\{(\mu_{air}\chi_{air} + \mu_{water}\chi_{water} + \mu_{oil}\chi_{oil})\left(\frac{\partial u_i}{\partial x_j} + \frac{\partial u_j}{\partial x_i}\right)\right\} \\
&+\rho\chi_L g_T\left(\frac{u_{T_1}}{\sqrt{|u_T|^2 + \varepsilon_g^2}}(T_1\chi_{oil})_i + \frac{u_{T_2}}{\sqrt{|u_T|^2 + \varepsilon_g^2}}(T_2\chi_{oil})_i\right) + \rho K_i \\
&-\chi_b \frac{\partial}{\partial x_j}\left(H \frac{\partial}{\partial x_j}\chi_{oil}\right) \cdot n_i^{oil} \quad \text{in}\, \Omega, \quad t > 0,
\end{aligned} \tag{7.16}$$

where

$$H = -\mu_{water}(k_{oil} + k_{water}) + (\mu_{water} - \mu_{oil})\, u_j^{oil} \cdot n_j. \tag{7.17}$$

The third term in (7.16) represents adhesion and sliding phenomena occurring at the interface between oil and sand. The fourth term is the dipole moment distribution along the reaction surface included in the bacteria zone, which contributes to satisfying the reactivity condition.

The phase equation is

$$\chi_{air} + \chi_{water} + \chi_{oil} = 1 \quad \text{in}\, \Omega, \quad t > 0. \tag{7.18}$$

The motion equation of free surface for air is

$$\frac{\partial \chi_{air}}{\partial t} + u_j \frac{\partial \chi_{air}}{\partial x_j} = 0 \quad \text{in}\, \Omega,\ t > 0. \tag{7.19}$$

The motion equation of free surface for oil is

$$\frac{\partial \chi_{oil}}{\partial t} + u_j^{oil} \frac{\partial \chi_{oil}}{\partial x_j} = k_{oil} \cdot |\nabla \chi_{oil}|\, \chi_b \quad \text{in}\, \Omega,\ t > 0, \tag{7.20}$$

Finally, the outer boundary conditions are prescribed as

$$\begin{cases} u_n = u_0 \sin\omega t, \\ u_T = 0, \end{cases} \text{on } \Gamma_I(t), \tag{7.21}$$

$$\begin{cases} u_n = 0, \\ \dfrac{\partial u_T}{\partial n} = 0, \end{cases} \text{on } \Gamma_S, \tag{7.22}$$

$$\begin{cases} u_n = 0, \\ u_T = 0, \end{cases} \text{on } \Gamma_N, \tag{7.23}$$

$$\begin{cases} \sigma_n = 0, \\ u_T = 0, \end{cases} \text{on } \Gamma_A. \tag{7.24}$$

Under suitable initial conditions for $\{u_i.\chi_{air}, \chi_{oil}\}$, (7.15) to (7.24) constitute a mathematical model representing the decomposition mechanism of oil.

7.7 Numerical results

Figures 7.2 and 7.2 shows the time-sequence of the oil decomposition process. The top figure shows the initial state in which air, oil and water in layers are at rest. A rectangle located near the center of the figure shows the bacteria habitat. We can see that the jump of velocity component normal to the boundary between oil and water occurrs while decomposition proceeds.

We observe (1) oil region reduction and water region increase corresponding to decomposition progress, (2) strong vorticity distribution near the decomposition surface.

Figure 7.3 shows the 3D case in the same situation. See the appendix which includes color figures. Colors in Fig. 7.3 indicate vorticity distribution on the reaction surface.

References

Duvaut G, Lions JL (1976) Inequalities in Mechanics and Physics, Springer-Verlag

Fujita H, Kawahara H, Kawarada H (1995) Distribution theoretic approach to fictitious domain method for Neumann problems. East-West J Numer Math 3:111–126

Glowinski R, Lions JL, Tremolieres R (1981) Numerical Analysis of Variational Inequalities, North-Holland publishing Company

Kawarada H, Fujita H, Suito H (1998) Wave motion breaking upon the shore. GAKUTO International Series Mathematical Sciences and Applications 11:145–159

Kawarada H, Suito H (1997) Numerical method for a free surface flow on the basis of the fictitious domain method. East-West J Numer Math 5:57–66

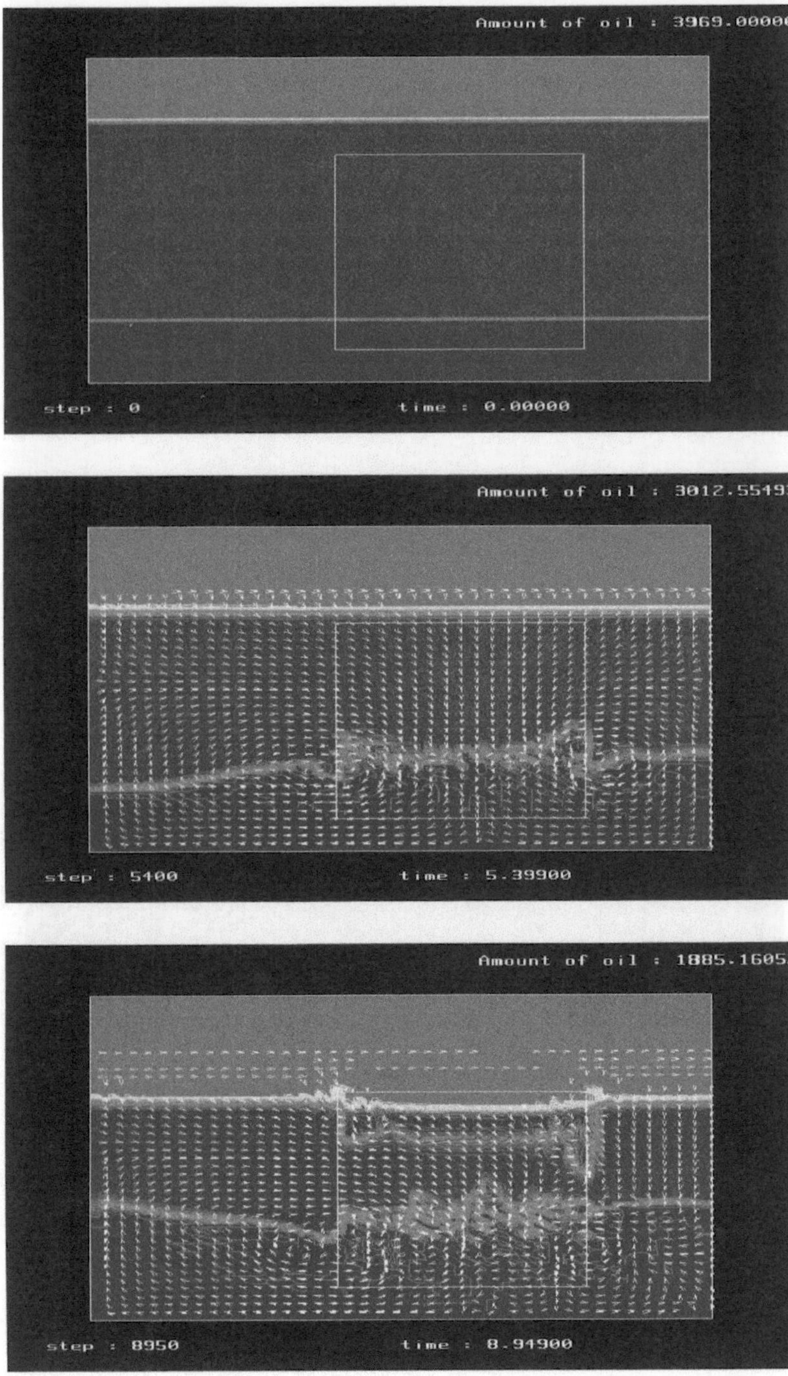

Fig. 7.2. Decomposition of oil in the 2D case

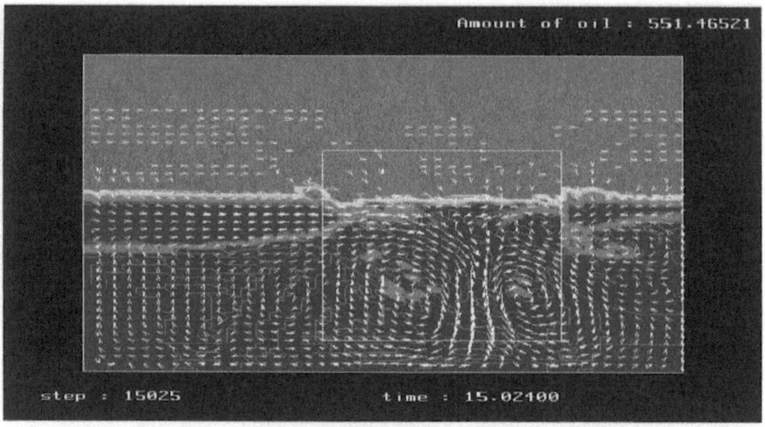

Fig. 7.2. Decomposition of oil in the 2D case (continued)

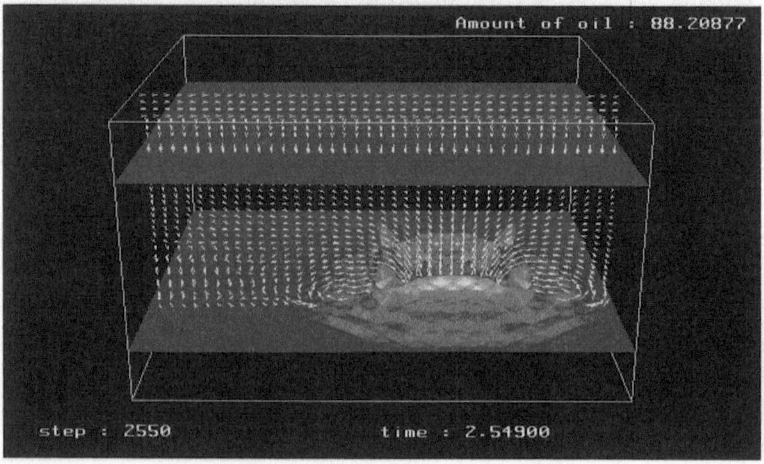

Fig. 7.3. Decomposition of oil in the 3D case

8 Breaking waves and ecosystem dynamics

8.1 Introduction

The role of waves at margins of a sandy beach or a tidal flat to sea water for determination of species survival in a coastal ecosystem was discussed in Section 2.3.3 from a viewpoint of environmental experiments. This means that different kinds of life appear under different habitat conditions such as with waves and without waves, i.e., the wave at the margin constitutes an important factor for determination of species composing coastal ecosystems.

Existence or nonexistence of green algae on the beach greatly influences macrobenthos in a tidal flat. If the beach is covered with green algae, seawater infiltration is hindered, bringing about shortage of oxygen and nutrients. Therefore, benthos suffer while surviving there. On the other hand, wave motion on the beach disposes algae. The reason for this phenomenon is considered as follows. Breaking waves bring fresh seawater into the seabed. Air bubbles are entrained through the action of breaking waves. Thus, a lot of oxygen is transported into the seabed. This is a main role of waves. The tide itself, on the other hand, does not contain air bubbles, but infiltrates seawater smoothly and gradually into the seabed: so the oxygen transport is relatively slight. It is supposed that benthos have their own sensor mechanism which detects the amount of oxygen brought by breaking waves; whereupon the benthos become active and consume nutrients around them. Consequently, the growth of algae is suppressed.

In this chapter, we interpret this experimental result by constructing a mathematical model for aeration with close relation to breaking waves and ecosystem dynamics. In this modeling, a phase field model to treat alloys and an optimization method based on genetic algorithms play important roles. The sensor peculiar to benthos combines both of them.

8.2 Mathematical modeling

8.2.1 Mathematical description of aeration due to breaking waves

Here, we follow mathematical formulation stated in Chapter 3. Furthermore, by referring to the concept utilized in the phase field model, the key of which is to use the double-well potential $G(\lambda)$ (λ: an order parameter), we try to model an aeration phenomenon due to breaking waves. The order parameter distinguishes phases of material. For example, $\lambda = 1$ indicates one phase and $\lambda = 0$ indicates another. Under such choices, $\lambda = 0$ and $\lambda = 1$ are stable equilibrium points and $\lambda = 1/2$ is an unstable one.

Here, let us regard the characteristic function χ_{air} used in multi-phase flow formulation as the order parameter. From now, let χ_{air} represent the occupation ratio of air per unit volume in Ω, which is defined by λ.

$$\lambda = \begin{cases} 0 & : \text{the state of water} \\ b \ \ (0 < b < 1) & : \text{the mixed state of air and water} \\ 1 & : \text{the state of air} \end{cases} \tag{8.1}$$

A sharply edged λ indicates a non-aerated state while a bluntly edged λ shows an aerated state.

Let us introduce the evolution equation for λ by use of the mass conservation law;

$$\frac{\partial \lambda}{\partial t} + \frac{\partial}{\partial x_j}(\lambda u_j) = -G'(\lambda), \quad \text{in } \Omega,\ t > 0, \tag{8.2}$$

where $G'(\cdot)$ means the derivative of $G(\cdot)$ and $\dfrac{\partial}{\partial x_j}(\lambda u_j)$ means $u_j \dfrac{\partial \lambda}{\partial x_j}$ because we assume $\dfrac{\partial u_j}{\partial x_j} = 0$ for $0 \le \lambda < 1$.

We propose this definition:

$$G(\lambda) = \begin{cases} \dfrac{h}{4}\lambda^2(\lambda - 1)^2 & ; 0 < \lambda < 1, \\ +\infty & ; \lambda \le 0,\ 1 \le \lambda. \end{cases} \tag{8.3}$$

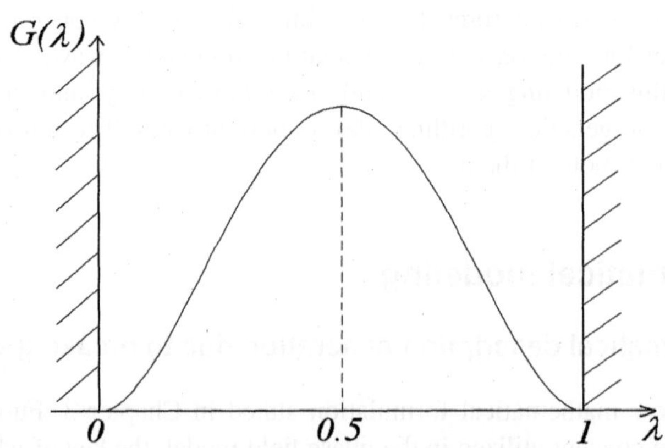

Fig. 8.1. Double well potential $G(\lambda)$

Addition of $-G'(\lambda)$ as a source control term in (8.2) brings about the following properties:

(1) The walls of $G(\lambda)$, $(\lambda \leq 0,\ \lambda \geq 1)$ physically avoid penetration of λ outside of $[0, 1]$.
(2) $-G'(\lambda)$ takes a non-zero value only in the narrow strip Ω_{strip} along the free surface. In the air side region $\Omega_{air} \cap \Omega_{strip}$, $-G'(\lambda)$ takes a positive value, meaning that it is the source. In the water side region $\Omega_{water} \cap \Omega_{strip}$, $-G'(\lambda)$ takes a negative value, meaning that it is the sink. The source increases the value of λ in Ω_{air}. On the other hand, the sink reduces the value of λ in Ω_{water}. In other words, $-G'(\lambda)$ has the effect of letting λ be sharply edged in Ω_{strip}. Furthermore, we note that $\int_{\Omega} G'(\lambda)dx$ is approximately zero.
(3) The value of $G(\lambda)$ defined near $\lambda = 1/2$, which we call the separating value, contributes to the unstable equilibrium at $\lambda = 1/2$. That is to say, if the separating value is sufficiently large, then λ becomes sharply edged, i.e., non-aerated state; if the separating value is sufficiently small, λ becomes bluntly edged, i.e., the aerated state.

Hypotheses

From a physical view point, we propose the following hypotheses:

(1) Strong vorticity distribution in a two-phase flow system reflects that the separating value is small.
(2) Weak vorticity distribution in a two-phase flow system reflects that the separating value is large.

Concrete representation of $h(\cdot)$

Define $h = h(|\omega|^2)$ (ω : vorticity) as a separating value. Following hypotheses 1 and 2, we assume $h(\xi) = \alpha\cdot \max\{1 - \xi/|\omega_0|\}$, ($\xi \in R^+, \alpha$ and ω_0 are positive constants.)

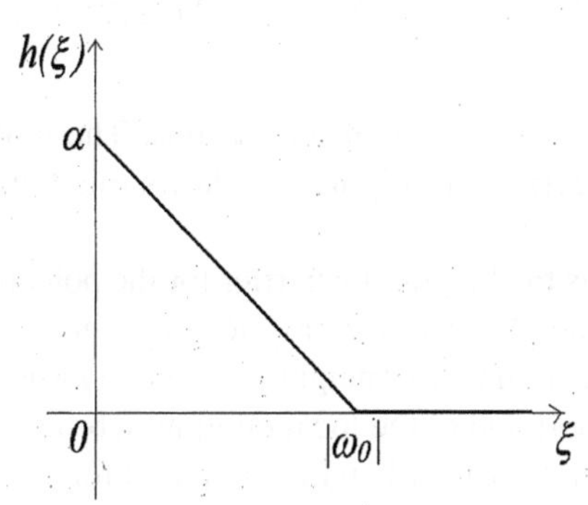

Fig. 8.2. Separating value $h(\xi)$

Remark 8.1

Vorticity distribution of the flow in this model corresponds to temperature distribution in a phase field model treating an alloy.

Remark 8.2

Let us note that governing equations for flow in the aerated state are defined under an assumption of extended stress tensor.

8.2.2 Modeling for ecosystem dynamics

We already know that presence or absence of waves causes stable dynamics due to suppression of algae or suffering of benthos due to thickening of green algae on the beach. (See Chapter 2.)

In this chapter, an idea to describe these phenomena is sought from genetic algorithms (GAs): (Goldberg 1989) An essential character of GAs consists in mathematical description of the concept involved in Darwinism. Now this description, equipped with mathematical formulation, comes home in order to represent ecosystem dynamics.

GAs to search for a global minimizer (maximizer) for a certain cost function were constructed by scattering a set of individuals in searching space and then removing them in the space by use of three fundamental operations of *selection, cross over*, and *mutation*, They promote transition between generations.

Of course, an aspect of multi-individuals to look for the global minimizer in GAs depends upon the cost function structure. If a cost function has a lot of local minimizers, multi-individual shall settle into the global minimizer passing through various paths, each of which includes some of local minimizers. In order to illustrate distribution of colonies in an ecosystem, we set up a lot of local minimizers that display a kind of ecological multi-stability.

As an example of a cost function, we choose the *Rastrigin function* to represent ecological stability:

$$f(\boldsymbol{x}) = N \cdot A + \sum_{i=1}^{N} (x_i^2 - A\cos(2\pi x_i)), \tag{8.4}$$

where $A = 10$ and N is a number of dimensions. The global minimizer of this function is $\boldsymbol{x} = \boldsymbol{0}$ and $f(\boldsymbol{0}) = 0$. Figure 8.3 shows this funcion with $N = 2$ and $\Omega = [-3.12, 3.12]^2$.

The value of A shows the height of a barrier for the potential to represent a local minimum. Low barrier height makes it easy to propagate individuals in searching space. On the other hand, high barrier height gives an opposite effect, i.e., traps some individuals in the neighborhood of the local minimum, bringing about a blockade of information. Then moderate barrier height is required for sound ecosystem dynamics.

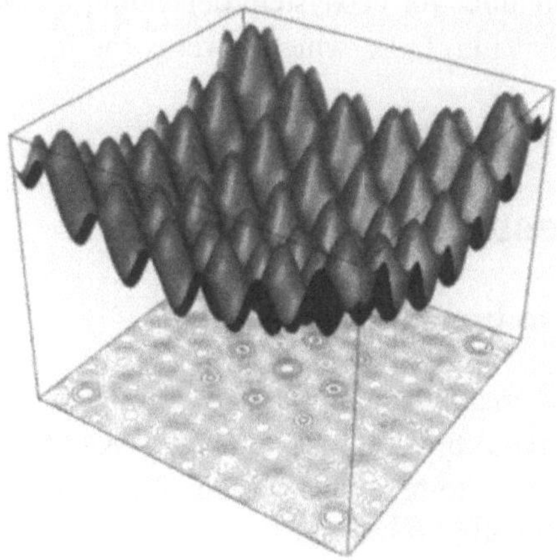

Fig. 8.3. 2D Rastrigin function

8.3 Coupling scheme between waves and ecosystem dynamics

Following the concept of sensors peculiar to benthos stated in the introduction, we develop the following.

- We scatter a set of sensors in the neighborhood of the beach.
- $\mathbf{S} = \{S_i\}_{i=1}^N$, where S_i is the i-th sensor.
- Let the value of order parameter λ at the i-th sensor setup point λ_i and define the response value of i-th sensor as

$$\Gamma_i = \begin{cases} 1, \text{ if } \dfrac{1}{2} \leq \lambda_i \leq 1, \\ 0, \text{ if } 0 \leq \lambda_i < \dfrac{1}{2}. \end{cases} \tag{8.5}$$

  The total response value of S is defined by $\Gamma = \sum_{i=1}^{N} \Gamma_i$.

  Let G be a set of parameters characterizing three fundamental elements in GAs. We define two types of G as:
  * $G^{(1)}$: A set giving normal function to GAs.
  * $G^{(2)}$: A set inducing functional disorder to GAs.
- How to put $G^{(i)}$ $(i = 1, 2)$ on the rearrangement operation in GAs.
  * If $\Gamma \geq \frac{N}{2}$, then GAs chooses $G^{(1)}$.
  * If $\Gamma < \frac{N}{2}$, then GAs chooses $G^{(2)}$.

Then, an aspect of dynamics for ecosystem depending on waves is determined by embedding $G^{(1)}$ or $G^{(2)}$ into GAs, which governs the rearrangement operation in GAs.

8.4 Numerical results

8.4.1 Aeration due to breaking waves

We numerically solved mathematical model stated in 8.2.1 in order to understand aeration mechanism due to breaking waves. Figure 8.4 shows that air bubbles roll in a run-up point of waves to breaking point along the surface of the beach, from which aerated air infiltrated into sea bed.

8.4.2 Simulation of ecosystem dynamics

Let many individuals scatter into search space at an initial state of GAs. After several operations of rearrangement in GAs, an aspect of dynamics to repeat the generation transfigures, while individuals change their positions. Some of them are trapped into local minimizers and escape from them. Gradually, they concentrate their positions into the global minimizer. These behaviors show stable ecosystem dynamics represented by group A in Fig. 8.5. On the other hand, group B in Fig. 8.5 show unstable dynamics. Characteristics of unstable dynamics are that individuals group together like crows, and then suddenly disappear as a mist, as seen in Fig. 8.5

- Group A : Mutation possibility = 0.01
- Group B : Mutation possibility = 0.1

- Size of population = 9
- Number of generations = 100

References

Kawarada H, Fujita H, Suito H (1998) Wave motion breaking upon the shore. GAKUTO International Series Mathematical Sciences and Applications 11:145–159

Goldberg F (1989) Genetic Algorithms in Search, Optimization, and Machine Learning, Addison-Wesley

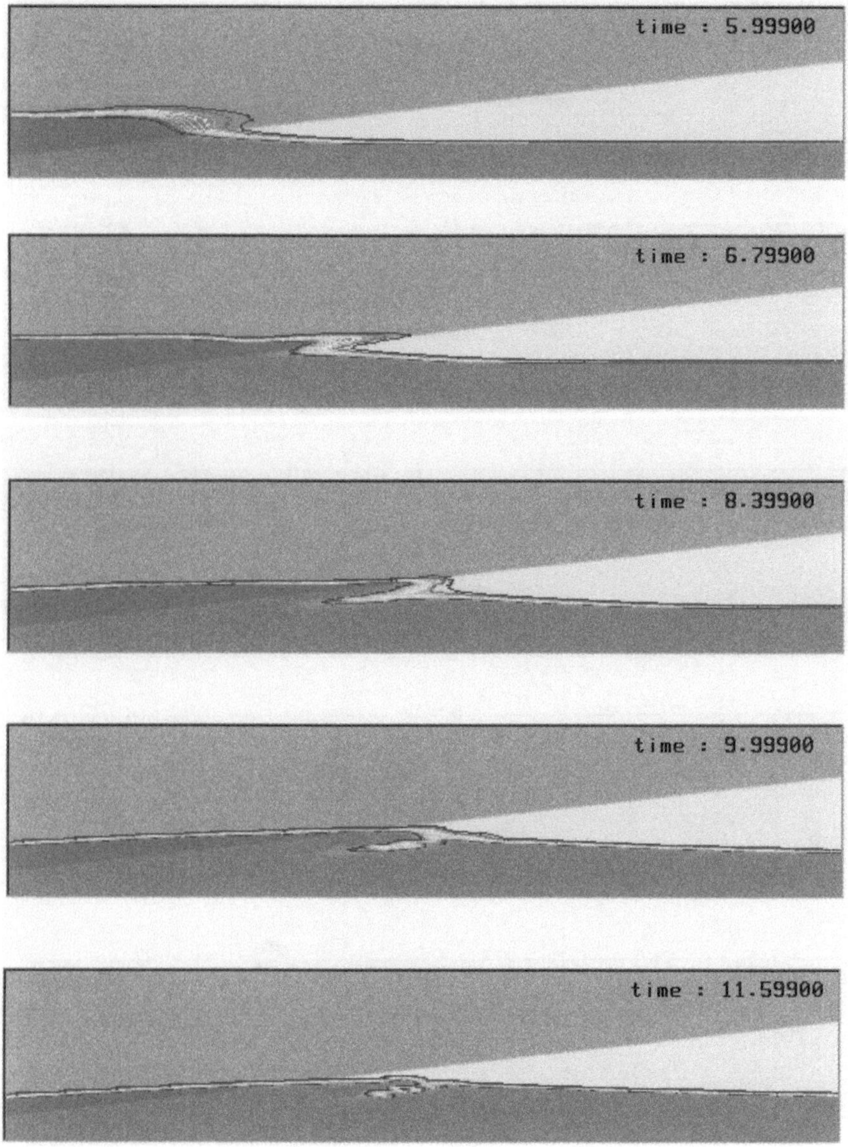

Fig. 8.4. Aeration

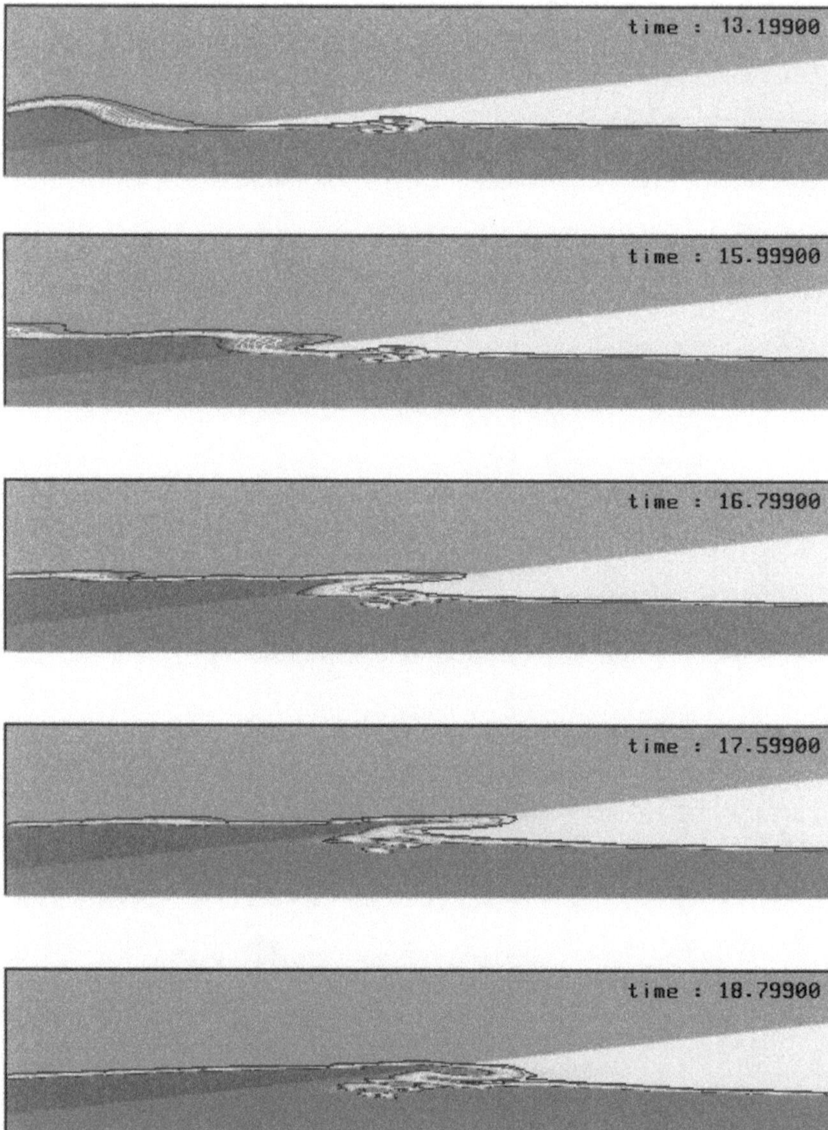

Fig. 8.4. Aeration (continued)

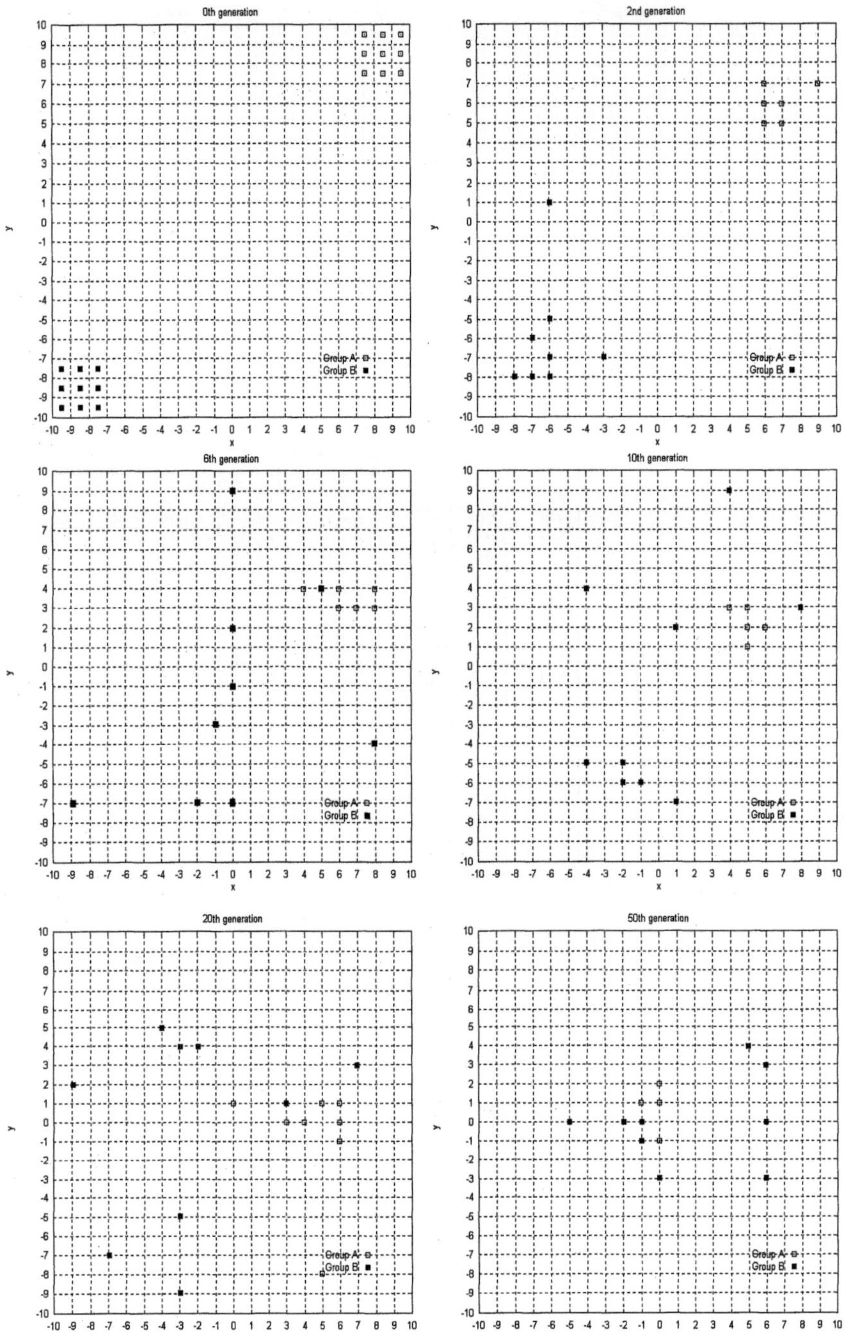

Fig. 8.5. Ecosystem dynamics

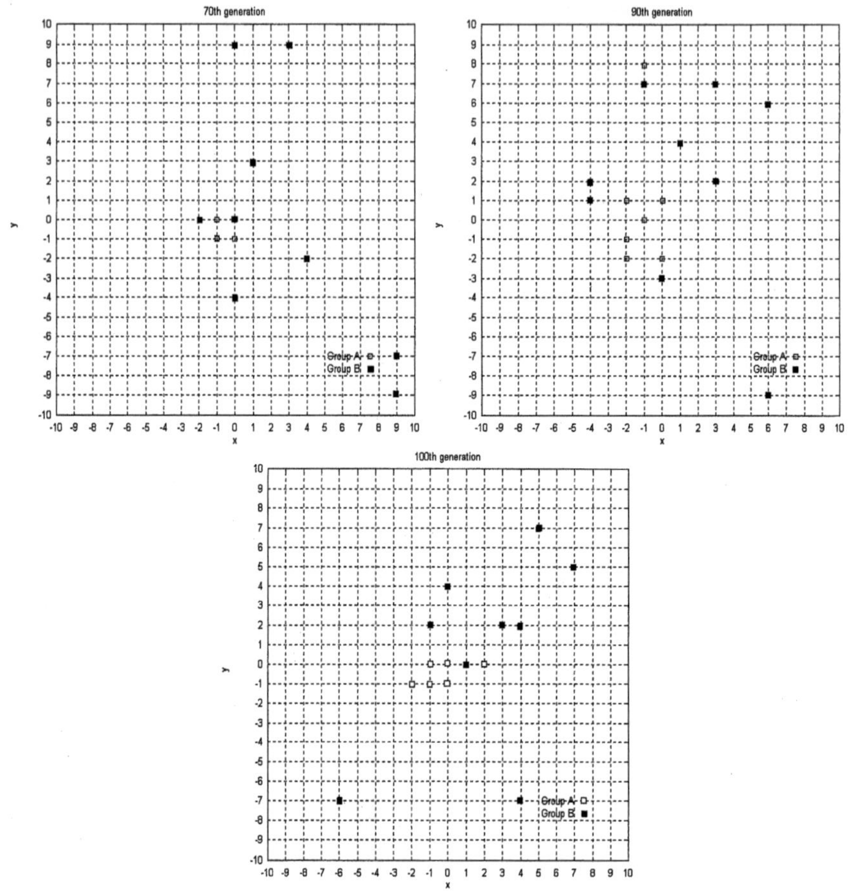

Fig. 8.5. Ecosystem dynamics (continued)

9 Methodologies for theoretical studies

9.1 Introduction

In this chapter, we explain the fundamental methodologies for theoretical studies stated in Chaps. 3, 5, 7 and 8, which correspond to facilities utilized in experimental studies in Chaps. 1, 2, 4 and 6. These methodologies consist of a mathematical part and a numerical one, which have been developed by the authors to promote the studies concerned. These methodologies are combined to obtain fruitful results *in the computer* which can be observed in the form of the many figures included in this book. They have contributed to deeper understanding of remarkable works of nature at environmental margins.

9.2 Mathematical methodologies

When building a mathematical model for multi-phase flow with interfacial interactions, the *distribution theoretic approach to multi-phase flow* is effectively used. This section explains this approach for the case of steady Stokes problem which does not lose its generality. On the other hand, smearing phenomena arising in the numerical free surface is a serious problem. A device to restrain this phenomenon is presented by referring to a phase field model with double-well potential.

9.2.1 Distribution theoretic approach to multi-phase flow

Let us note that penetration, adhesion and decomposition phenomena discussed in Chaps. 3, 5, and 7 are described by using Delta-measure supported on Γ_B and the interface between different phases. These phenomena are represented by jump boundary conditions for the Dirichlet type (the reactivity condition) and the Neumann type (the friction condition). In order to introduce such jump conditions into a unified model for a multi-phase flow system, we have developed a *distribution theoretic approach to multi-phase flow*.

For simplicity, we discuss the case of a steady Stokes problem defined in $\Omega \subset R^3$. Ω is divided into subdomains Ω_1 and Ω_2, whose interface is denoted by Γ. (Fig. 9.1.) The two-phase Stokes problem (S_1) with jump boundary conditions on Γ is defined as follows:

$$(S_1)\quad \begin{cases} -\Delta u^{(1)} + \nabla p^{(1)} = f^{(1)} & \text{in } \Omega_1, \\ \text{div } u^{(1)} = 0 & \text{in } \Omega_1, \\ -\Delta u^{(2)} + \nabla p^{(2)} = f^{(2)} & \text{in } \Omega_2, \\ \text{div } u^{(2)} = 0 & \text{in } \Omega_2, \\ [\sigma_n] = \sigma_n^{(1)} - \sigma_n^{(2)} = a & \text{on } \Gamma, \\ [\sigma_{T_j}] = \sigma_{T_j}^{(1)} - \sigma_{T_j}^{(2)} = b_j \ (j = 1, 2) & \text{on } \Gamma, \\ [u_n] = u_n^{(1)} - u_n^{(2)} = c & \text{on } \Gamma, \\ [u_{T_j}] = u_{T_j}^{(1)} - u_{T_j}^{(2)} = d_j \ (j = 1, 2) & \text{on } \Gamma, \\ u^{(1)} = 0, & \text{on } \partial\Omega. \end{cases} \tag{9.1}$$

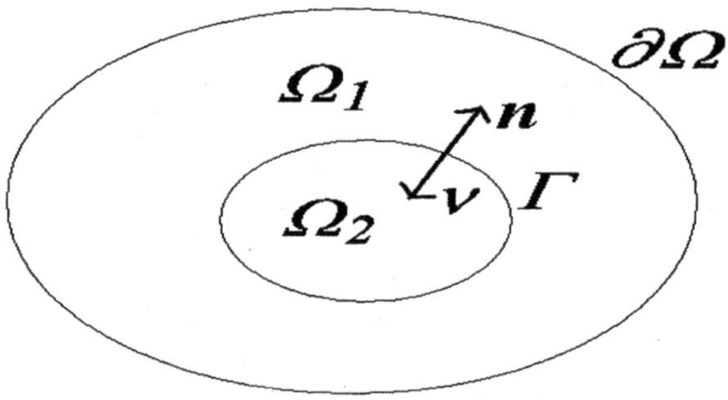

Fig. 9.1. Geometry of Stokes problem

Let χ be the characteristic function of Ω_2 in Ω and assume a, b_j, c, and d_j to be constants. Then, (S_1) is settled into a single equation (S_2) in the following way:

$$(S_2)\quad \begin{cases} -\Delta u + \nabla p + a \cdot \nabla\chi + b_1 \cdot \Sigma_1\chi + b_2 \cdot \Sigma_2\chi & \\ \quad +c \cdot \boldsymbol{n} \cdot \Delta\chi + d_1 \cdot T_1 \cdot \Delta\chi + d_2 \cdot T_2 \cdot \Delta\chi = f, & \text{in } \Omega, \\ \text{div}\boldsymbol{u} = 0, & \text{in } \Omega\backslash\Gamma, \\ u = 0, & \text{on } \partial\Omega, \end{cases} \tag{9.2}$$

where

$$f = (1 - \chi)f^{(1)} + \chi f^{(2)}, \tag{9.3}$$

and

$$\nabla\chi = \left(\frac{\partial\chi}{\partial x_1}, \frac{\partial\chi}{\partial x_2}, \frac{\partial\chi}{\partial x_3}\right), \tag{9.4}$$

$$\Sigma_1\chi = \left(\frac{\partial\chi}{\partial x_2}, -\frac{\partial\chi}{\partial x_1}, 0\right), \tag{9.5}$$

$$\Sigma_2\chi = \frac{\left(\dfrac{\partial\chi}{\partial x_1}\dfrac{\partial\chi}{\partial x_3}, \dfrac{\partial\chi}{\partial x_2}\dfrac{\partial\chi}{\partial x_3}, \left(\dfrac{\partial\chi}{\partial x_3}\right)^2 - |\nabla\chi|^2\right)}{\sqrt{\left(\dfrac{\partial\chi}{\partial x_1}\right)^2 + \left(\dfrac{\partial\chi}{\partial x_2}\right)^2}}, \tag{9.6}$$

$$\boldsymbol{n} = -\frac{\nabla\chi}{|\nabla\chi|}, \quad T_1 = \frac{\Sigma_1\chi}{|\Sigma_1\chi|}, \quad T_2 = \frac{\Sigma_2\chi}{|\Sigma_2\chi|}. \tag{9.7}$$

Here, $\dfrac{\partial\chi}{\partial x_j}$ $(j = 1, 2, 3)$ is a distribution supported on Γ, multiplication and division of them are defined by convolution of distribution with support. Summarizing the above mentioned facts, we have

Theorem 9.1
(S_1) is equivalent to (S_2).

Remark 9.1
Surface distribution of the dipole moment with normal direction to the surface brings about the jump of the normal component of velocity for fluid flowing across the interface.

Proof of Theorem 9.1 The case of the jump condition for the Dirichlet type
Let us consider the problem $(Pr)_c$ to generate $[u_n] = c$. For simplicity, let us put $c = 1$. This yields

$$(Pr)_c \begin{cases} \triangle u - \nabla p = \tilde{n}\triangle\chi & \text{in } \Omega, \\ \operatorname{div} u = 0 & \text{in } \Omega, \\ u = 0 & \text{on } \partial\Omega, \end{cases} \tag{9.8}$$

where $\triangle\chi = \frac{\partial}{\partial\nu}\delta(\Gamma)$ is a dipole moment, $\Gamma = \partial\Omega_2$ is smooth and $\tilde{n}$ is smoothly extended into R^3 so as to satisfy $\left.\dfrac{\partial\tilde{n}_j}{\partial\nu}\right|_{\Gamma} = 0$ $(j = 1, 2, 3)$.

[1st step] Potential flow is defined to satisfy

$$(Pr)_{c1} \begin{cases} \triangle U - \nabla P = \tilde{n}\cdot\triangle\chi & \text{in } R^3, \\ \operatorname{div} U = 0 & \text{in } R^3. \end{cases} \tag{9.9}$$

We define fundamental solution: $E^k = \left\{E_j^k\right\}_{j=1}^3$ $(k = 1, 2, 3)$ s.t.

$$\begin{cases} \triangle E^k - \nabla q^k = \delta(x - y)\cdot e_k & \text{in } R^3, \\ \operatorname{div} E^k = 0 & \text{in } R^3, \end{cases} \tag{9.10}$$

where e^k is an unit vector of the kth axis $(k = 1, 2, 3)$ and $x = (x_1, x_2, x_3)$, $y = (y_1, y_2, y_3)$.

$$\begin{cases} E_j^k(x,y) = -\dfrac{\delta_j^k}{4\pi\,|x-y|} + \dfrac{(x_k-y_k)(x_j-y_j)}{|x-y|^3} \\ \qquad\qquad = \delta_j^k E(x-y) + F_j^k(x-y) \in L_y^1(R^3), \\ q^k(x,y) = \dfrac{\partial}{\partial x_k} E(x-y) \in L_y^1(R^3). \end{cases} \tag{9.11}$$

Potential $\{U, P\}$ is represented as

$$\begin{cases} U(x) = E * \tilde{n}\triangle\chi \ \text{ in } \mathcal{D}'(R^3), \\ P(x) = q * \tilde{n}\triangle\chi \ \text{ in } \mathcal{D}'(R^3), \end{cases} \tag{9.12}$$

where $*$ represents a convolution. In (9.12), $\mathcal{D}'$ means a dual space of $\mathcal{D}$, which is a set of infinitely differentiable functions with compact support. This means that the Laplacian $\triangle$ is defined in the sense of distribution.

$$\begin{aligned} U_j(x) &= \int_{\Omega_2} \triangle_y \left(E_j^k(x-y)\cdot\tilde{n}_k(y)\right) d\Omega_y, \\ &= \int_{\Omega_2} \mathrm{div}_y\nabla_y \left(E_j^k(x-y)\cdot\tilde{n}_k(y)\right) d\Omega_y, \\ &= \int_{\Gamma} \frac{\partial}{\partial n_y} E_j^k(x-y)\cdot\tilde{n}_k(y) d\sigma_y, \\ &= \int_{\Gamma} \frac{\partial}{\partial n_y} E(x-y)\cdot\tilde{n}_j(y) d\sigma_y \\ &\quad + \int_{\Gamma} \frac{\partial}{\partial n_y} F_j^k(x-y)\cdot\tilde{n}_k(y) d\sigma_y. \end{aligned} \tag{9.13}$$

The second term of the right hand side does not contribute to make the jump because

$$F_j^k = E * \frac{\partial^2}{\partial x_k \partial x_j}\delta = \frac{\partial^2}{\partial x_k \partial x_j} E. \tag{9.14}$$

The first term is a double layer potential. Then, U satisfies

$$[U] = U|_{\Gamma_+} - U|_{\Gamma_-} = n, \tag{9.15}$$

$$n\cdot[U] = n\cdot n = 1. \tag{9.16}$$

Similarly, P satisfies

$$P(x) = \int_{\Omega_2} \triangle\left\{q^k(x,y)\cdot\tilde{n}_k(y)\right\} d\Omega_y, \tag{9.17}$$

$$= \int_{\Gamma} \frac{\partial}{\partial n_y} q^k(x,y)\cdot n_k(y) d\sigma_y, \tag{9.18}$$

$$= \frac{\partial}{\partial x_k} \int_\Gamma n_k(y) \frac{\partial}{\partial n_y} E(x-y) d\sigma_y. \tag{9.19}$$

Then there holds

$$[P] = P|_{\Gamma_+} - P|_{\Gamma_-} = 0. \tag{9.20}$$

Let us note that $\boldsymbol{U}$ and P are real analytic in $R^3 \backslash \Gamma$.

[2nd step]

Let $u = U + \tilde{u}$ and $p = P + \tilde{p}$. Then $(\tilde{u}, \tilde{p})$ satisfies

$$(Pr)_{c2} \begin{cases} \Delta\tilde{u} - \nabla\tilde{p} = 0 \text{ in } \Omega, \\ \text{div}\tilde{u} = 0 \qquad \text{in } \Omega, \\ \tilde{u} = -U \qquad \in \left\{H^{\frac{1}{2}}(\partial\Omega)\right\}^2, \end{cases} \tag{9.21}$$

where $\int_\Omega \text{div} U \, d\Omega = \int_\Gamma U \cdot n \, d\sigma = 0.$

Generally, Sobolev space $H^m = \{f \in \mathcal{D}', \ \partial^\alpha f \in L^2, \ (|\alpha| \leq m)\}$ where m is an integer and $\partial^\alpha f$ means a differentiation in the sense of distribution. Sobolev space H^s ($s \in \mathbf{R}$) is defined by generalizing the integer case via Fourier transformation. See (Schwartz, 1961) for details.

Note that there exists a unique solution $\{\tilde{u}, \tilde{p}\}$ for $(Pr)_{c2}$ satisfying

(1) $\tilde{u} \in \left\{H^1(\Omega)\right\}^2$,
(2) $\tilde{p} \in L^2(\Omega) \backslash R$.

Then $n \cdot [u] = n\,[U] + n\,[\tilde{u}] = n\,[U] = 1$ in $\left\{H^{\frac{1}{2}}(\Gamma)\right\}^2$.

Theorem 9.2

The solution of $(Pr)_c$ *satisfies that* $n \cdot [u] = 1$ *in* $\left\{H^{\frac{1}{2}}(\Gamma)\right\}^2$.

The case of the jump condition for the type of $[\sigma_n] = a$

Let us consider the problem $(Pr)_a$ to generate $[\sigma_n] = a(= 1)$:

$$(Pr)_a \begin{cases} \Delta u - \nabla p = \nabla\chi \text{ in } \Omega, \\ \text{div} u = 0 \qquad \text{in } \Omega, \\ u = 0 \qquad \text{on } \partial\Omega. \end{cases} \tag{9.22}$$

[1st step] Potential flow is defined to satisfy

$$(Pr)_{a1} \begin{cases} \Delta U - \nabla P = \nabla\chi & \text{in } R^3, \\ \text{div}\, U = 0 & \text{in } R^3. \end{cases} \tag{9.23}$$

So, clearly

$$\begin{cases} U(x) = 0 & \text{in } R^3, \\ P(x) = -\chi & \text{in } R^3 \end{cases} \tag{9.24}$$

is the solution of $(Pr)_{a1}$.

[2nd step]

$u = U$ and $p = P$ satisfies $(Pr)_a$. Then we have

Theorem 9.3

$$\left[\frac{\partial u_n}{\partial n} - p\right] = 1 \text{ in } \left\{H^{-\frac{1}{2}}(\Gamma)\right\}^2. \tag{9.25}$$

The case of the jump condition for the type of $[\sigma_{T_l}] = b_l (= 1)(l = 1, 2)$

Let us consider the problem $(Pr)_b$ to generate $[\sigma_{T_l}] = b_l$:

$$(Pr)_b \begin{cases} \Delta u - \nabla p = \Sigma_l \chi \text{ in } \Omega, \\ \operatorname{div} u = 0 & \text{in } \Omega, \\ u = 0 & \text{on } \partial\Omega. \end{cases} \tag{9.26}$$

[1st step] Potential flow is defined to satisfy

$$(Pr)_{b1} \begin{cases} \Delta U - \nabla P = \Sigma_l \chi & \text{in } R^3, \\ \operatorname{div} U = 0 & \text{in } R^3. \end{cases} \tag{9.27}$$

Then we have

$$\begin{aligned} U_j(x) &= \int_{R^3} E_j^k(x-y) \cdot t_k^{(l)}(y) \cdot \delta(y - \Gamma)\, dy \\ &= \int_{\Gamma} E_j^k(x-y) \cdot t_k^{(l)}(y)\, d\sigma_y, \end{aligned} \tag{9.28}$$

$$t_k^{(l)} = \frac{(\Sigma_l \chi)_k}{|\Sigma_l \chi|}, \tag{9.29}$$

$$\delta_\Gamma = |\Sigma_l \chi|, \tag{9.30}$$

and

$$U_{\tilde{T}}(x) = \int_{\Gamma} E_j^k(x-y) \cdot t_j^{(l)}(x) \cdot t_k^{(l)}(y)\, d\sigma_y. \tag{9.31}$$

From the above, it follows that

$$(\tilde{n} \cdot \nabla) U_{\tilde{T}} = -\int_{\Gamma} \frac{\partial}{\partial n_y} E(x-y) \cdot t_j^{(l)}(x) \cdot t_j^{(l)}(y)\, d\sigma_y + \cdots. \tag{9.32}$$

By use of no contribution of F_j^k to make the jump, we have

$$\left[\frac{\partial}{\partial n} U_T\right] = -1. \tag{9.33}$$

[2nd step]

Repeating similar arguments as before, we have

Theorem 9.4

$$\left[\frac{\partial u_T}{\partial n}\right] = -1 \text{ in } \left\{H^{-\frac{1}{2}}(\Gamma)\right\}^2. \tag{9.34}$$

The Stokes equation with variable viscosity In this section, we shall address the Stokes equation with a variable viscosity in place of the one with constant viscosity discussed in previous Sects. as

$$\begin{cases} \dfrac{\partial}{\partial x_j}\left(\nu(x)\dfrac{\partial u_i}{\partial x_j}\right) - \dfrac{\partial p}{\partial x_i} = f_i \text{ in } \Omega \text{ for } i = 1, 2, \\ \dfrac{\partial u_j}{\partial x_j} = 0 \qquad\qquad\qquad\qquad\quad \text{in } \Omega, \end{cases} \tag{9.35}$$

where $\nu(x) = \nu_1(x)(1-\chi) + \nu_2(x)\chi$ and $\nu_j(x) \in C(\Omega_j)$ $(j = 1, 2)$.

According to the replacement of the Laplacian by the divergence form, the jumped boundary conditions defined on Γ should be modified as

$$[\sigma_n] = \nu_1 \frac{\partial u_{1n}}{\partial n} - p_1 - \left(\nu_2 \frac{\partial u_{2n}}{\partial n} - p_2\right), \tag{9.36}$$

$$[\sigma_T] = \nu_1 \frac{\partial u_{1T}}{\partial n} - \nu_2 \frac{\partial u_{2T}}{\partial n}, \tag{9.37}$$

$$[u_n] = \nu_1 u_{1n} - \nu_2 u_{2n}, \tag{9.38}$$

$$[u_T] = \nu_1 u_{1T} - \nu_2 u_{2T}. \tag{9.39}$$

However, if the relation $u_{1n} - u_{2n} = c$ on Γ is required in place of $[u_n] = c$, then $n \cdot \triangle\chi$ should be replaced with $n \cdot \frac{\partial}{\partial x_j}\left(H \frac{\partial}{\partial x_j}\chi\right)$. In fact,

$$\nu_1 u_{1n} - \nu_2 u_{2n} = \nu_1(u_{1n} - u_{2n}) + (\nu_1 - \nu_2)u_{2n} \text{ on } \Gamma. \tag{9.40}$$

Then H should be defined on Γ in the following way:

$$H = c\nu_1 + (\nu_1 - \nu_2)u_{2n} \text{ on } \Gamma. \tag{9.41}$$

The additional terms except $\boldsymbol{n} \cdot \triangle\chi$ in (S_2) brings about the same jumps as in the statement of Theorem 9.1.

Remark 9.2 *A distribution theoretic approach for Stokes equation with variable viscosity is proved by treating the transmission problem defined on Γ under the weak formulation of the equation, that will be shown in the succeeding paper.*

Remark 9.3 *In the case of time dependent Stokes equations, we can show the same result as obtained in the steady case.*

9.2.2 Anti-smearing device for numerical free surface

Various types of methods have been proposed (Hirt and Nichols, 1981; Sussman et al. 1994) to deal with free surface problems. They intend to avoid numerical dissipation and/or dispersion caused by discretization of an evolution equation for free surfaces. In our solution algorithm, a simpler technique based on Gibbs free-energy density in the phase field model is used. In this technique, a source control term G' is added to equations of free surfaces to avoid numerical dissipation:

$$\frac{\partial \chi_{air}}{\partial t} + \frac{\partial}{\partial x_j}(\chi_{air} u_j) = -G'(\chi_{air}), \tag{9.42}$$

and

$$\frac{\partial \chi_{oil}}{\partial t} + \frac{\partial}{\partial x_j}(\chi_{oil} u_j) = -G'(\chi_{oil}), \tag{9.43}$$

where $G'(\chi)$ is the derivative of a double-well potential $G(\chi)$, which is defined as follows:

$$G(\chi) = \begin{cases} +\infty & (\chi < 0), \\ 16h\chi^2(\chi - 1)^2 & (0 < \chi < 1), \\ +\infty & (1 < \chi). \end{cases} \tag{9.44}$$

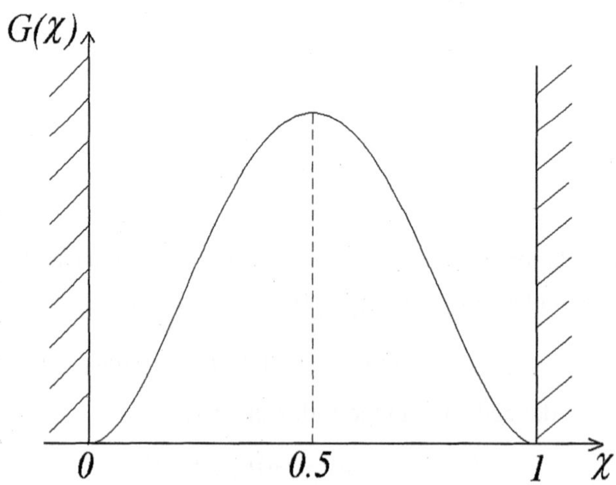

Fig. 9.2. $G(\chi)$

The walls of $G(\chi)$ ($\chi \leq 0$, $1 \leq \chi$) prevent χ from deviating from the interval $[0, 1]$ to the outside. The term $-G'(\chi)$ takes a non-zero value only in the narrow strip $\widetilde{\Omega} = \widetilde{\Omega}_{(+)} \cup \widetilde{\Omega}_{(-)} \cup \{\boldsymbol{x} \,|\chi(\boldsymbol{x}) = 0.5\}$ along the free surface, where $\widetilde{\Omega}_{(+)} = \{\boldsymbol{x} \,|\, 0.5 < \chi(\boldsymbol{x}) < 1\}$ and $\widetilde{\Omega}_{(-)} = \{\boldsymbol{x} \,|\, 0 < \chi(\boldsymbol{x}) < 0.5\}$. In $\widetilde{\Omega}_{(-)}$, $-G'(\chi)$ takes a negative value, which implies a sink of χ. In $\widetilde{\Omega}_{(+)}$, $-G'(\chi)$ takes a positive value, which implies a source of χ. Consequently, $-G'(\chi)$ has the effect of letting χ be sharply edged. The coefficient h in Eq. (9.44) is a "wall height" at $\chi = 0.5$, which is determined according to the slope of the characteristic function χ in the vicinity of $\{\mathbf{x}|\chi(\mathbf{x}) = 0.5\}$. This study determines h as

$$h = \widetilde{h}\left(\frac{1}{d \cdot \Delta} - |\nabla\chi|\right)^{+}, \tag{9.45}$$

where Δ is a typical grid spacing: for example, $\Delta = \max(\Delta x, \Delta y)$. Here, $\widetilde{h}$ is a positive constant and $(w)^{+}$ means the positive part of $w \in \mathbf{R}$, i.e., $(w)^{+} = \dfrac{w + |w|}{2}$. This means that when width $\widetilde{d}$ of a narrow strip (i.e., $\{\mathbf{x}|0 < \chi(\mathbf{x}) < 1\}$) exceeds $d \cdot \Delta$, the source control term G' begins to work because $|\nabla\chi| \simeq \dfrac{1}{\widetilde{d}}$. Numerical experiments show that this rather simple technique works well for the sake of avoiding numerical dissipation of free surfaces.

9.3 Numerical methodologies

This section plays the role of a bridge to link together mathematical models and observations *in the computer*. The statement of 9.3.2 points out the importance of an algorithm in the numerical method, without which it is impossible to obtain stable numerical solutions.

9.3.1 Discretized model for the total system

We have presented several unified equations for total flow systems in Chaps. 3, 5, and 7. Here, for example, the numerical procedure to solve unified equations (5.17)-(5.21) defined in Chap. 5 is summarized as follows;

Step 1: Solve the Poisson equation due to Helmholtz decomposition to obtain the pressure field in Ω.

$$\triangle p^{n+1} = \frac{\partial Q_j^n}{\partial x_j} + \frac{1}{\Delta t}\rho^n \frac{\partial u_j^n}{\partial x_j}, \tag{9.46}$$

where

$$\begin{aligned} Q_i = {} & \rho^n\left(-u_j^n \frac{\partial u_i^n}{\partial x_j} - c\chi_L u_i^n + K_i\right) \\ & + \frac{\partial}{\partial x_j}\left[\left\{\mu_{air}\chi_{air}^n + \mu_{water}\chi_{water}^n + \left(\mu_{oil}^{\infty} + \frac{\mu_{oil}^{0} - \mu_{oil}^{\infty}}{1 + (KD_{II}^n)^m}\right)\chi_{oil}^n\right\}\left(\frac{\partial u_i^n}{\partial x_j}\right)\right] \\ & - \rho^n \chi_L g_T \left[\frac{\{u_j^n(T_1\chi_{oil}^n)_j\}(T_1\chi_{oil}^n)_i}{\sqrt{(u_j^n(T_1\chi_{oil}^n)_j)^2 + \varepsilon_s^2}} + \frac{\{u_j^n(T_2\chi_{oil}^n)_j\}(T_2\chi_{oil}^n)_i}{\sqrt{(u_j^n(T_2\chi_{oil}^n)_j)^2 + \varepsilon_s^2}}\right]. \end{aligned} \tag{9.47}$$

Superscript n means the nth time step, i.e., $t = n\Delta t$.

In Eq. (9.46), the incompressibility condition at $n+1$ time step is assumed. The GPBi-CG method (Zhang, 1998) with a preconditioner due to incomplete LU factorization is employed for solving (9.46).

Step 2: Set initial guess of u_i and χ_α for iteration at the $n+1$ time step.

$$k = 0, \tag{9.48}$$
$$u_i^{n+1,k} = u_i^n, \tag{9.49}$$
$$\chi_{oil}^{n+1,k} = \chi_{oil}^n, \tag{9.50}$$
$$\chi_{air}^{n+1,k} = \chi_{air}^n, \tag{9.51}$$

where second superscript k means the iteration step number at the n-th time step.

Step 3: Iterate the unified equation of motion and the evolution equations for free surfaces in total domain Ω using pressure field p^{n+1} obtained in Step 1.

$$\chi_{water}^{n+1,k} = 1 - \chi_{air}^{n+1,k} - \chi_{oil}^{n+1,k} \tag{9.52}$$
$$\rho^{n+1,k} = \rho_{air}\chi_{air}^{n+1,k} + \rho_{water}\chi_{water}^{n+1,k} + \rho_{oil}\chi_{oil}^{n+1,k} \tag{9.53}$$
$$\mu_{oil}^{n+1,k} = \mu_{oil}^{\infty} + \frac{\mu_{oil}^0 - \mu_{oil}^{\infty}}{1 + \left(KD_{\Pi}^{n+1,k}\right)^m} \tag{9.54}$$

$$\begin{aligned}
&\frac{u_i^{n+1,k+1} - u_i^n}{\Delta t} + u_j^n \frac{\partial u_i^{n+1,k+1}}{\partial x_j} = -\frac{1}{\rho^{n+1,k}}\frac{\partial p^{n+1}}{\partial x_i} + K_i - c\chi_L u_i^{n+1,k+1} \\
&+\frac{1}{\rho^{n+1,k}}\frac{\partial}{\partial x_j}\left\{\left(\mu_{air}\chi_{air}^{n+1,k} + \mu_{water}\chi_{water}^{n+1,k} + \mu_{oil}^{n+1}\chi_{oil}^{n+1,k}\right)\frac{\partial u_i^{n+1,k+1}}{\partial x_j}\right\} \\
&-\chi_L \frac{g_T}{\rho^{n+1,k}}\left[\frac{\left\{u_j^{n+1,k+1}(T_1\chi_{oil}^{n+1,k})_j\right\}(T_1\chi_{oil}^{n+1,k})_i}{\sqrt{(u_j^{n+1,k}(T_1\chi_{oil}^{n+1,k})_j)^2 + \varepsilon_g^2}}\right. \\
&\qquad\left.\frac{\left\{u_j^{n+1,k+1}(T_2\chi_{oil}^{n+1,k})_j\right\}(T_2\chi_{oil}^{n+1,k})_i}{\sqrt{(u_j^{n+1,k}(T_2\chi_{oil}^{n+1,k})_j)^2 + \varepsilon_g^2}}\right]
\end{aligned} \tag{9.55}$$

$$\frac{\chi_{oil}^{n+1,k+1} - \chi_{oil}^n}{\Delta t} + \frac{\partial}{\partial x_j}\left(\chi_{oil}^{n+1,k+1}u_j^{n+1,k+1}\right) = -G'(\chi_{oil}^{n+1,k}) \tag{9.56}$$
$$\frac{\chi_{air}^{n+1,k+1} - \chi_{air}^n}{\Delta t} + \frac{\partial}{\partial x_j}\left(\chi_{air}^{n+1,k+1}u_j^{n+1,k+1}\right) = -G'(\chi_{air}^{n+1,k}) \tag{9.57}$$

Step 4: If residuals satisfy convergence criteria

$$\max\left|u_i^{n+1,k+1} - u_i^{n+1,k}\right| < \varepsilon_u, \tag{9.58}$$
$$\max\left|\chi_{oil}^{n+1,k+1} - \chi_{oil}^{n+1,k}\right| < \varepsilon_{oil}, \tag{9.59}$$
$$\max\left|\chi_{air}^{n+1,k+1} - \chi_{air}^{n+1,k}\right| < \varepsilon_{air}, \tag{9.60}$$

then stop iteration at the n-th time step, increment n and return to Step 2.

If residuals don't satisfy convergence criteria, then increment k and return to Step 3.

Boundary conditions imposed on $\partial\Omega$ are:

$$\begin{cases} \chi_{water}\,(u_1 - A \sin \omega t) + \chi_{air}\dfrac{\partial u_1}{\partial x_1} = 0, \\ u_2 = 0, \\ u_3 = 0, \end{cases} \quad \text{on } \Gamma_{left}, \tag{9.61}$$

$$\begin{cases} \chi_{water} u_1 + \chi_{air}\dfrac{\partial u_1}{\partial x_1} = 0, \\ u_2 = 0, \\ u_3 = 0, \end{cases} \quad \text{on } \Gamma_{right}, \tag{9.62}$$

$$\begin{cases} \dfrac{\partial u_1}{\partial x_2} = 0, \\ u_2 = 0, \\ \dfrac{\partial u_3}{\partial x_2} = 0, \end{cases} \quad \text{on } \Gamma_{bottom}, \tag{9.63}$$

$$\begin{cases} u_1 = 0, \\ \dfrac{\partial u_2}{\partial x_2} = 0, \\ u_3 = 0, \end{cases} \quad \text{on } \Gamma_{top}, \tag{9.64}$$

$$\begin{cases} \dfrac{\partial u_1}{\partial x_3} = 0, \\ \dfrac{\partial u_2}{\partial x_3} = 0, \\ u_3 = 0, \end{cases} \quad \text{on } \Gamma_z. \tag{9.65}$$

where Γ_{left}, Γ_{right}, Γ_{top}, and Γ_{bottom} are left, right, top, and bottom boundaries, respectively. (See Fig. 5.1, which is a projection of computational domain to $x - y$ plane.) Lateral boundary Γ_z is perpendicular to the z-axis.

All spatial derivatives except for advection terms in (9.55) can be approximated by second order central differences. The advection term in (9.55) can be approximated by the third order upwind scheme: (Kawamura and Kuwahara, 1984)

9.3.2 Two-phase free surface flow with large density difference

If we try to solve numerically the multi-phase flow problem stated in Chapters 3, 5, 7 and 8, we follow the numerical method described in 9.3.1. Then we certainly encounter an unstable phenomenon arising in the motion of the interface between air and water (or oil) phases. We have noticed that this instability may be due to large difference of density between air and water (or oil). In this subsection, the cause of this phenomenon is clarified and a method for its remedy is presented.

Notations

u	: Velocity field
p	: Pressure
$\{\Omega_a, \Omega_w\}$	: Domains occupied with air and water
$\Gamma = \overline{\Omega_a} \cap \overline{\Omega_w}$	: Free surface
$\Omega = \Omega_a \cup \Gamma \cup \Omega_w$	: Total domain
$\partial\Omega$	: Boundary of Ω
$\{\chi_a, \chi_w\}$	: Characteristic functions of Ω_a and Ω_w
g	: Acceleration of gravity
$\{\rho_a, \rho_w\}$	: Densities of air and water
$\{\nu_a, \nu_w\}$	: Kinematic viscosities of air and water
$D_{ij}(u)$	: Strain tensor
e_k	: Unit vector of the kth axis ($k = 1, 2, 3$)

Two-phase free surface flow problem We shall deal with the following simple problem for two phase-free surface flow, defined in $\mathbf{R}^2$, which does not lose its essential character to induce instability:

$$(P)\begin{cases} \dfrac{\partial u_i}{\partial t} + u_j \dfrac{\partial u_i}{\partial x_j} = -\dfrac{1}{\rho}\dfrac{\partial p}{\partial x_i} + \dfrac{\partial}{\partial x_j}(\nu D_{ij}) - g\delta_{i,3}, & \text{in } \Omega,\ t > 0, \\ \dfrac{\partial u_j}{\partial x_j} = 0, & \text{in } \Omega,\ t > 0, \\ \dfrac{\partial \chi_w}{\partial t} + u_j \dfrac{\partial \chi_w}{\partial x_j} = 0, & \text{in } \Omega,\ t > 0, \\ u_i = 0, & \text{on } \partial\Omega,\ t > 0, \\ u_i|_{t=0} = 0, & \text{in } \Omega, \\ \chi_w|_{t=0} = \chi_0, & \text{in } \Omega, \end{cases} \tag{9.66}$$

where $\rho = \rho_a\chi_a + \rho_w\chi_w$ $(\dfrac{\rho_a}{\rho_w} = \dfrac{1}{1000})$, $\nu = \nu_a\chi_a + \nu_w\chi_w$ and $\chi_a + \chi_w = 1$.

The cause of instability When solving (P) under a curved initial state for an interface, fluid motion is expected to stop after a long time and the interface approaches the horizontal asymptotically. The stationery state for (P) should be described as $\boldsymbol{u} = 0$ and $\nabla p + \rho g e_2 = 0$, under which the interface stays at the horizontal. If we rewrite $\nabla p + \rho g e_2 = 0$ as $\nabla(p + \rho_a g x_2) + (\rho_w - \rho_a)\chi_w g\, \mathbf{e}_2 = 0$, then the equilibrium condition in the discretized version becomes

$$(\mathrm{E})_h = \nabla_h p_h + (\rho_w - \rho_a)\chi_h \cdot g \cdot \mathbf{e}_2 \simeq 0, \tag{9.67}$$

where χ_h means the discretized χ_w. A discretized one for Navier-Stokes equation in (P) is

$$\frac{\mathbf{u}_h^{n+1} - \mathbf{u}_h^n}{\Delta t} = A(\rho_h^n, \nu_h^n ; \mathbf{u}_h^n) - \nabla p_h^n - \rho_h^n \cdot g e_2, \tag{9.68}$$

where $A(\rho_h^n, \nu_h^n ; \mathbf{u}_h^n)$ includes both discretized viscosity and convection terms. Here we define

$$\mathbf{V}_h^{n+1} = \mathbf{u}_h^n + \Delta t\{A(\rho_h^n, \nu_h^n; \mathbf{u}_h^n) - \rho_h^n \cdot g\mathbf{e}_2\}, \tag{9.69}$$

which we call an auxiliary vector field. We should note that the incompressibility condition div $\boldsymbol{u} = 0$ in (P) is satisfied by use of Helmholtz decomposition of an auxiliary vector field $\mathbf{V}_h^{n+1}$; this means $\mathbf{V}_h^{n+1} = \Delta t \nabla_h p_h^{n+1} + \mathbf{u}_h^{n+1}$, where p_h^{n+1} is scalar potential and $\boldsymbol{u}_h^{n+1} = \mathrm{rot}_h \boldsymbol{E}_h$) ($\boldsymbol{E}_h$: vector potential).

Let Δ_h be the discrete Laplacian based on a 5 points formula. Then p_h is the solution of the pressure Poisson equation:

$$\begin{cases} \Delta_h p_h = \mathrm{div}_h \mathbf{V}_h & \text{in } \Omega_\mathrm{h}, \\ \dfrac{\partial p_h}{\partial n} = \mathbf{V}_h \cdot \mathbf{n}_h & \text{on } \partial\Omega_\mathrm{h}, \end{cases} \tag{9.70}$$

where n_h is the unit outward normal vector defined on $\partial\Omega_h$ and $\mathbf{V}_h = \mathbf{V}_h^0 + (\rho_w - \rho_a)\chi_h \mathbf{V}_h^1$.

- χ_h means the discretized interface
- $\mathbf{V}_h^1 \simeq -g \cdot \mathbf{e}_2$ near the horizontal
- $\Delta t = 1$

Inclusion of χ_h in V_h gives a crucial effect to interface instability.

Does $(\rho_w - \rho_a)\chi_h \cdot g \cdot \mathbf{e}_2$ reappear in $\nabla_h p_h$ after Helmholtz decomposition? The answer is no, because χ_h deforms into χ_h^* after Helmholtz decomposition. What is the difference between χ_h and χ_h^* ? Asymptotic form of $\nabla_h p_h$ near the horizontal is represented by

$$\nabla_h p_h = -(\rho_w - \rho_a) \cdot g \cdot (n_h^1 \cdot n_h^2 \mathbf{e}_1 + (n_h^2)^2 \mathbf{e}_2) \cdot \chi_h^* + \mathbf{R}_h \tag{9.71}$$

near the horizontal, by an effective use of the discretized fundamental solution, and the orthogonal curvilinear coordinate system. In (9.71), $\mathbf{R}_h$ is the residual part, $\nabla_h \chi_h = (D_h^1 \chi_h, D_h^2 \chi_h)$, and $\mathbf{n_h} = \dfrac{\nabla_h \chi_h}{|\nabla_h \chi_h|} = (n_h^1, n_h^2)$ at each mesh point. It should be noted that χ does not change in the continuous version.

1st step. The simplified problem

Let us consider (9.70) for the simple case that $\mathbf{V}_h = \chi_h \cdot \mathbf{e}_2$, which does not lose an essential character, and Ω_h is an infinite strip domain put in parallel with the horizontal in $\mathbf{R}^2$. Then the pressure Poisson equation reduces to the one-dimensional one, as

$$\begin{cases} \dfrac{p_{i-1} - 2p_i + p_{i+1}}{h^2} = (D_h^R \chi_h^R)_i = \dfrac{\chi_{i+1} - \chi_{i-1}}{2h}, \quad (i = 0, 1, \cdots, N), \\ p_0 = 0, \\ \dfrac{p_N - p_{N-1}}{h} = 0. \end{cases} \tag{9.72}$$

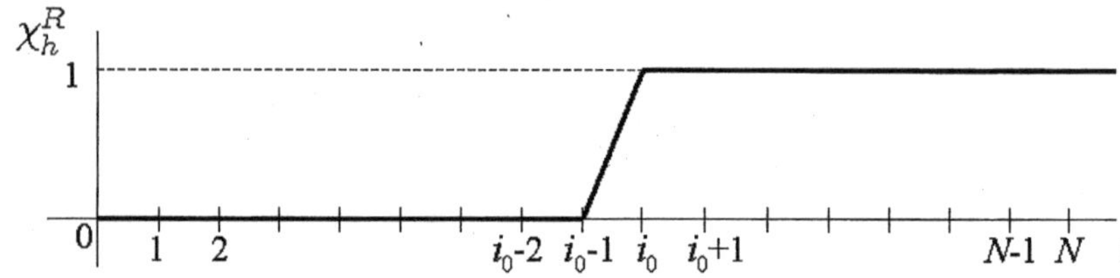

Fig. 9.3. χ_h

D_h^R and χ_h^R in (9.72) mean that they are defined by central differences due to regular meshes.

The solution p_h of (9.71) is represented by

$$(p_h)_i = \sum_{j=0}^{N} G_{h:i,j}(D_h^R \chi_h^R)_j \cdot h, \ (i = 1, 2, \cdots, N-1). \tag{9.73}$$

Using discrete Green's function,

$$G_{h:i,j} = \begin{cases} jh - 1, & (i \leq j), \\ ih - 1, & (j < i). \end{cases} \tag{9.74}$$

Then, the irrotational part for $\mathbf{V}_h = \chi_h \boldsymbol{e}_2$ is represented by

$$\begin{aligned} (D_h^R p_h)_i &= \sum_{j=0}^{N} (D_h^R G_h)_{i,j} (D_h^R \chi_h^R)_j \cdot h \\ &= \sum_{j=1}^{N-1} \delta_{h:i,j}^R (\chi_h^R)_j \cdot h, \ (i = 1, 2, \cdots, N-1) \end{aligned} \tag{9.75}$$

where

$$\delta_{h:i,j}^R = \begin{cases} 0, & j \geq i+2, \\ \dfrac{1}{4h}, & j = i+1, \\ \dfrac{1}{2h}, & j = i, \\ \dfrac{1}{4h}, & j = i-1, \\ 0, & j \leq i-2. \end{cases} \tag{9.76}$$

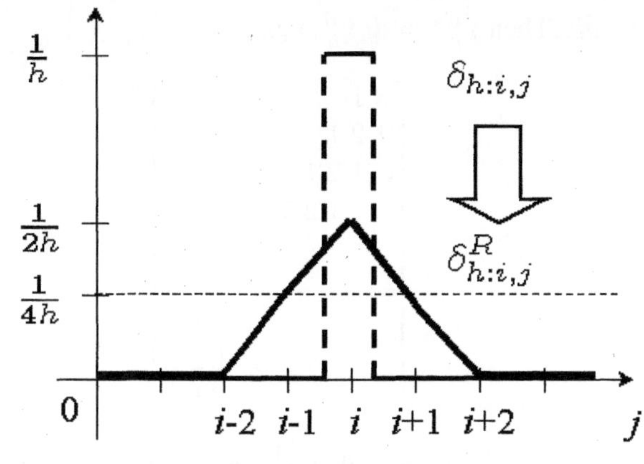

Fig. 9.4. δ_h^R and δ_h

$\delta^R_{h:i,j}$ implies some kind of relaxation for discrete Delta-measure $\delta_{h:i,j}$. Then define

$$
\begin{aligned}
(\chi_h^{R*})_i = (D_h^R p_h)_i &= \sum_{j=1}^{N-1} \delta^R_{h:i,j}(\chi_h^R)_j \cdot h \\
&= \frac{1}{4}\{(\chi_h^R)_{i-1} + 2(\chi_h^R)_i + (\chi_h^R)_{i+1}\} \\
&\quad (i = 1, 2, \cdots, N-1).
\end{aligned} \tag{9.77}
$$

We should note the difference between χ_h^R and χ_h^{R*} on $i = i_0 - 1$ and $i = i_0$, which

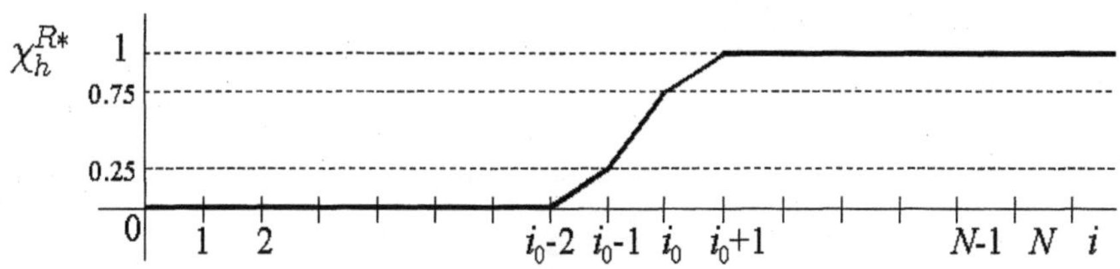

Fig. 9.5. χ_h^R

means the deformation of χ_h^R after Helmholtz decomposition.

Define $d_h = h\delta_h^R$. Then $\chi_h^{R*} = d_h \chi_h^R$, i.e.,

$$\chi_h^{R*} = \frac{1}{4}\begin{bmatrix} 3\,1 & & & & & & & \\ 1\,2\,1 & & & & & & & \\ & 1\,2\,1 & & & & & & \\ & & 1\,2\,1 & & & & & \\ & & & \cdots & & & & \\ & & & & \cdots & & & \\ & & & & & 1\,2\,1 & & \\ & & & & & & 1\,2\,1 & \\ & & & & & & & 1\,3 \end{bmatrix}. \tag{9.78}$$

On the other hand, we should note that for staggered mesh,

$$\begin{aligned}(\chi_h^{S*})_{i+\frac{1}{2}} = (D_h^S p_h)_{i+\frac{1}{2}} &= \sum_{j=0}^{N} (D_h^S G_h)_{i+\frac{1}{2},j} (D_h^S \chi_h^S)_j \cdot h \\ &= \sum_{j=1}^{N-1} \delta^S_{h:i+\frac{1}{2},j+\frac{1}{2}} \chi^S_{h:j+\frac{1}{2}} \cdot h \\ &= (\chi_h^S)_{i+\frac{1}{2}},\end{aligned} \tag{9.79}$$

where

$$(D_h^S p_h)_{i+\frac{1}{2}} = \frac{p_{i+1} - p_i}{h}, \tag{9.80}$$

$$(D_h^S \chi_h^S)_i = \frac{\chi_{i+\frac{1}{2}} - \chi_{i-\frac{1}{2}}}{h} \tag{9.81}$$

at each mesh point.

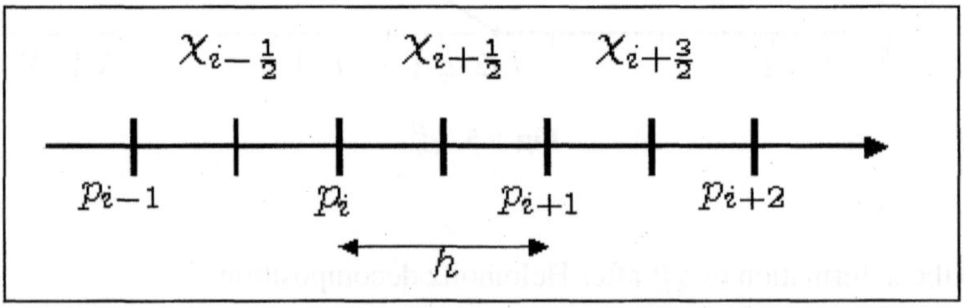

Fig. 9.6. Staggered mesh

Equation (9.79) indicates non-deformation of χ_h^S with use of a staggered mesh.

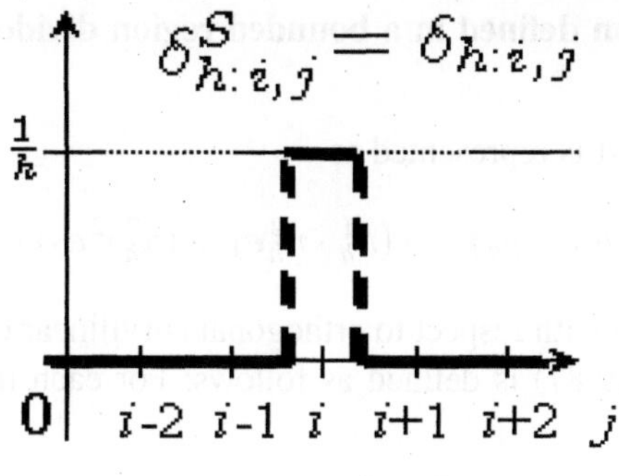

Fig. 9.7. Delta measure on staggered mesh

2nd step. The problem defined in a bounded region devided by the horizontal interface

Solve the pressure Poisson equation.

$$\begin{cases} \Delta_h p_h = \mathrm{div}_h \mathbf{V}_h \ \ \mathrm{in}\Omega_h \\ \dfrac{\partial p_h}{\partial n} = \boldsymbol{E}_h \cdot \boldsymbol{n}_h \ \mathrm{on}\partial\Omega_h \end{cases} \tag{9.82}$$

Then we have the irrotational part,

$$\nabla_h p_h = -(\rho_w - \rho_a) \cdot g \cdot \boldsymbol{e}_2 \cdot \chi^*_{0h} + \boldsymbol{R}_h, \tag{9.83}$$

near the horizontal. Here we used the discrete fundamental solution,

$$\boldsymbol{F}_{h:i,k;j,l} = \frac{1}{2\pi} \log \left[\left\{ (i-k)^2 + (j-l)^2 \right\} h^2 \right]^{-1/2}, \tag{9.84}$$

which satisfies

$$-\Delta_h \boldsymbol{F}_{h:i,k;j,l} = \frac{1}{h^2} \delta_{i,k} \cdot \delta_{j,l} + \mathrm{O}(h^2). \tag{9.85}$$

χ^*_{0h} is defined as

$$\chi^*_{0h} = \begin{bmatrix} d_h & & & \\ & d_h & & \\ & & \cdot & \\ & & & \cdot \\ & & & & d_h \end{bmatrix} \chi_{0h}, \tag{9.86}$$

where χ_{0h} means the discretized χ for the problem concerned.

3rd step. The problem defined in a bounded region divided by the curved interface

The irrotational part is represented as

$$\nabla_h p_h = -(\rho_w - \rho_a) \cdot g \cdot (n_h^1 \cdot n_h^2 \boldsymbol{e}_1 + (n_h^2)^2 \boldsymbol{e}_2) \cdot \chi_h^* + \boldsymbol{R}_h \qquad (9.87)$$

by transforming (9.83) with respect to orthogonal curvilinear coordinates.

Then term χ_h^* in (9.87) is defined as follows: For each mesh point $P \in \Omega_h$, define

$$\chi_h^*(\mathrm{P}) = \frac{1}{4}\{\chi_h(\mathrm{Q}) + 2\chi_h(\mathrm{P}) + \chi_h(\mathrm{R})\}. \qquad (9.88)$$

Above, $\chi_h(\mathrm{Q}) = \sum_{j=1}^{4} \alpha_j \chi_h(\mathrm{Q}_j)$, $\mathrm{Q}_4 = \mathrm{P}$, and α_j is an interpolating weight ($\sum_{j=1}^{4} \alpha_j = 1$, Fig. 9.8); $\chi_h(R)$ is defined in a similar way as above.

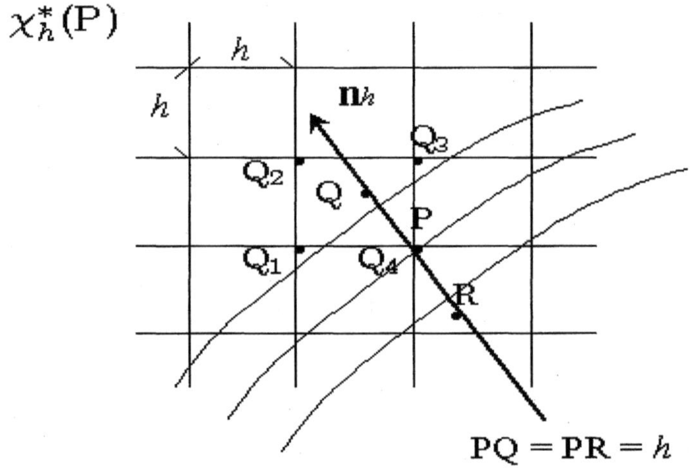

Fig. 9.8. Interpolation

Then we have $\chi_h^* = D_h(\boldsymbol{\alpha})\chi_h$, where

$$D_h = D_h(\alpha) = \frac{1}{4}\begin{bmatrix} D_{0h} & L_h^T & & & & \\ \cdot & \cdot & \cdot & & & \\ L_h & D_{0h} & L_h^T & & & \\ & L_h & D_{0h} & L_h^T & & \\ & & & \cdot & \cdot & \cdot \\ & & & L_h & D_{0h} & L_h^T \\ & & & & L_h & D_{0h} \end{bmatrix} \qquad (9.89)$$

$$D_{0h} = D_{0h}(\alpha) =$$

$$\begin{bmatrix} \cdot & \cdot & & & & \\ & \alpha_3(P_k) & 2+2\alpha_4(P_k) & \alpha_3(P_k) & & \\ & & \alpha_3(P_{k+1}) & 2+2\alpha_4(P_{k+1}) & \alpha_3(P_{k+1}) & \\ & & & & \cdot & \cdot & \cdot \end{bmatrix} \tag{9.90}$$

$$L_h = L_h(\alpha) = \begin{bmatrix} \cdot & \cdot & & & \\ & \alpha_1(P_k) & \alpha_2(P_k) & & \\ & & \alpha_1(P_{k+1}) & \alpha_2(P_{k+1}) & \\ & & & \cdot & \cdot \end{bmatrix} \tag{9.91}$$

Violation of Equilibrium Condition If we substitute (9.87) into the equilibrium condition (9.67), then we have

$$|(E)_h| = (\rho_w - \rho_a)|\chi_h^* - \chi_h| \cdot g \simeq \frac{1}{4}(\rho_w - \rho_a)g \tag{9.92}$$

near the horizontal.

Equation (9.92) gives rise to large apparent force ($\simeq \frac{1}{4}(\rho_w - \rho_a)g \cdot e_2$) acting on fluids near the stationary state due to large density difference between ρ_w and ρ_a. It should be noted that this apparent force does not disappear for any small $h(> 0)$. This is the cause of free surface instability in the neighborhood of the stationary state.

Remedy for instability (M1) Modification of $(E)_h$ by replacing χ_h by χ_h^*

Let us modify $(E)_h$ by replacing χ_h in the term of gravity force by χ_h^*:

$$(E^*)_h = \nabla_h p_h + (\rho_w - \rho_a)\chi_h^* \cdot g \cdot e_2 = 0. \tag{9.93}$$

Then solve the discretized Navier-Stokes equations after modification of χ_h into χ_h^* in step 2 of the solution algorithm. In fact, we found from repeated numerical experiments that harmonic average mapping is effective:

$$H_h = \frac{1}{2}(D_{h,(\alpha=(0,0,1,0))} + D_{h,(\alpha=(1,0,0,0))} + A_h), \tag{9.94}$$

where A_h is the arithmetic mean mapping with respect to 8 points which surround mesh point P. In practical computation, H_h is used in place of $D_h(\alpha)$.

$$H_h = \frac{1}{16} \begin{bmatrix} \cdot\ \cdot & \cdots & \cdots & & \\ \cdots & \cdots & \cdots & & \\ 1\,2\,1 & 2\,4\,2 & 1\,2\,1 & & \\ & 1\,2\,1 & 2\,4\,2 & 1\,2\,1 & \\ & \cdots & \cdots & \cdot\ \cdot & \\ & & \cdots & \cdots & \cdot \end{bmatrix} \tag{9.95}$$

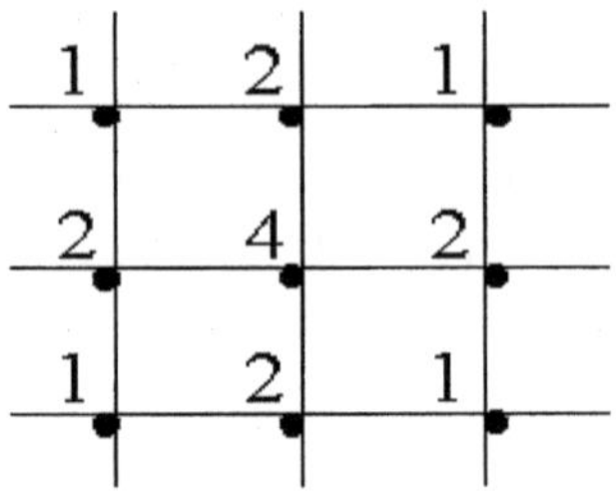

Fig. 9.9. H_h

(M2) Modification of $(E)_h$ by replacing $\nabla_h p_h$ by $(D_h(\boldsymbol{\alpha}))^{-1}\nabla_h p_h$

$$(\mathrm{E}^{**})_h = (D_h(\boldsymbol{\alpha}))^{-1}\nabla_h p_h + (\rho_w - \rho_a)\chi_h \cdot g \cdot \boldsymbol{e}_2 = 0. \tag{9.96}$$

Another way to remedy instability is to replace $\nabla_h p_h$ in discretized Navier-Stokes equations by $(D_h(\boldsymbol{\alpha}))^{-1}\nabla_h p_h$ in step 2 of the solution algorithm; that seems to be better than replacing χ_h by χ_h^* from a physical point of view, but it requires complicated procedures.

We conclude that modifications **(M1)** and **(M2)** improve the numerical method based on the regular mesh for two-phase free surface flow with large density difference. The instability arizen in the concerned computation could not be cleaned away until the way mentioned in this section was found.

References

Harlow FH, Welch JE (1965) Numerical calculation of time-dependent viscous incompressible flow of fluid with free surface. Phys Fluids 8:2182–2189

Hirt CW, Nichols BD (1981) Volume of fluid (VOF) method for the dynamics of free boundaries. J Comput Phys 39:201–225

Kawamura T, Kuwahara K (1984) Computation of high Reynolds number flow around a circular cylinder with surface roughness. AIAA paper 84-0340

Kawarada H, Fujita H, Suito H (1998) Wave motion breaking upon the shore. GAKUTO International Series Mathematical Sciences and Applications 11:145–159

Kawarada H, Suito H (1997) Numerical method for a free surface flow on the basis of the fictitious domain method. East-West J Numer Math 5(1):57–66

Schwartz L (1961) Méthodes Mathématiques pour les Sciences Physiques, Editions Scientifiques Hermann, Paris

Sussman M, Smereka P, Osher S (1994) A level set approach for computing solutions to incompressible two-phase flow. J Comput Phys 114:146–159

Zhang SL (1998) GPBi-CG: Generalized product-type methods based on Bi-CG for solving nonsymmetric linear systems. SIAM J Sci Comput 18(2):537–551

Appendix

This appendix includes color figures whose numbers are corresponding to the ones in the text.

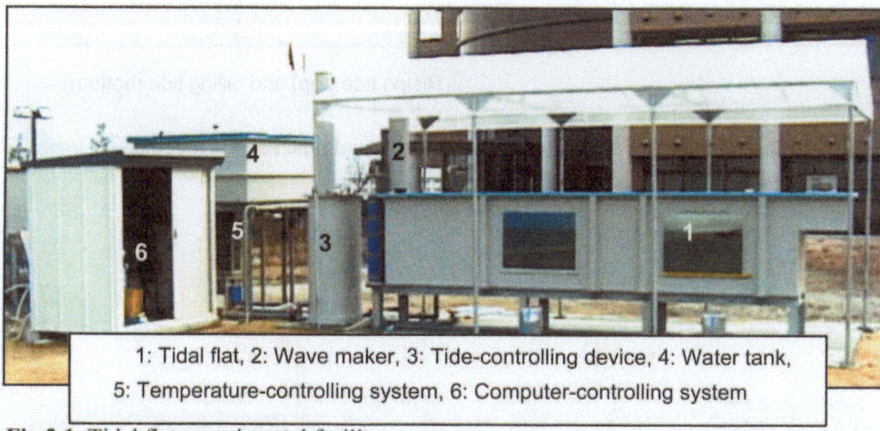

Fig.2.1. Tidal flat experimental facility

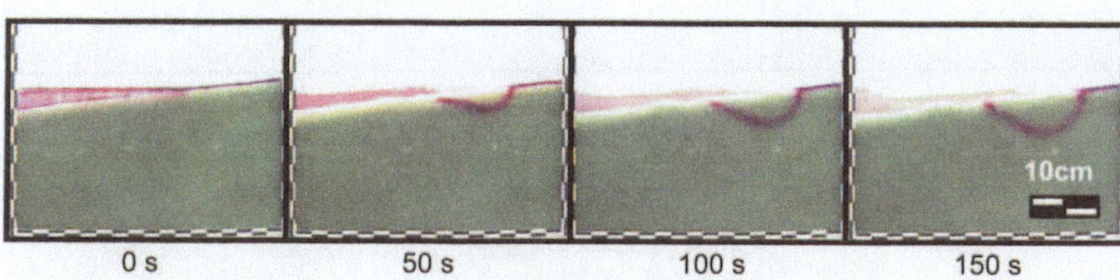

Fig. 2.2. Infiltration behavior of seawater due to wave action

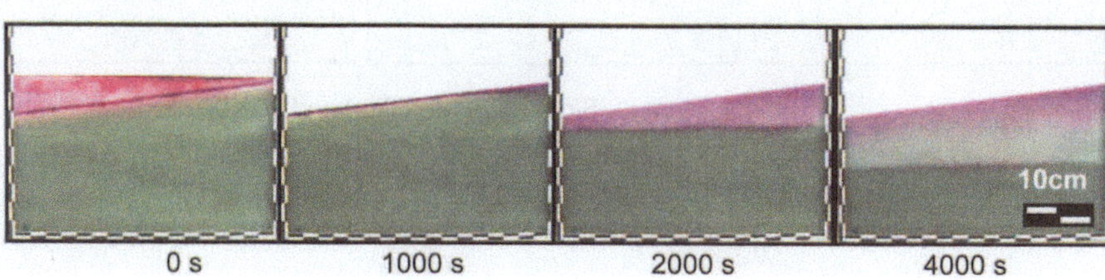

Fig. 2.4. Infiltration behavior of seawater due to tidal fluctuation

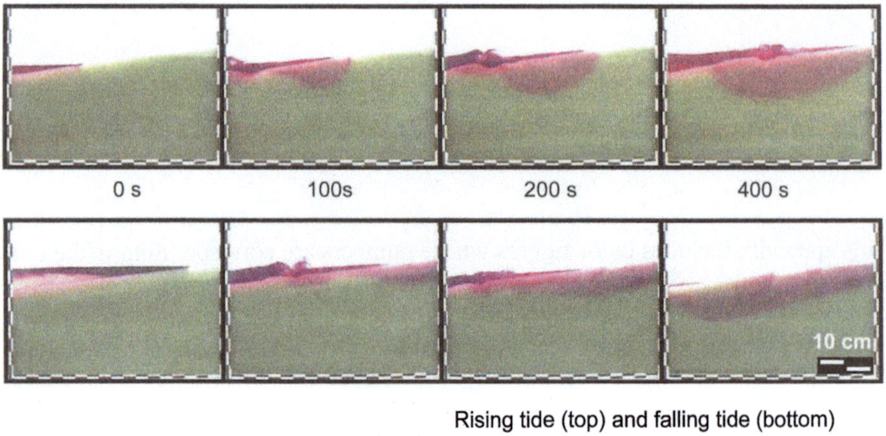

Fig. 2.5. Infiltration behavior of seawater due to wave and tidal actions together

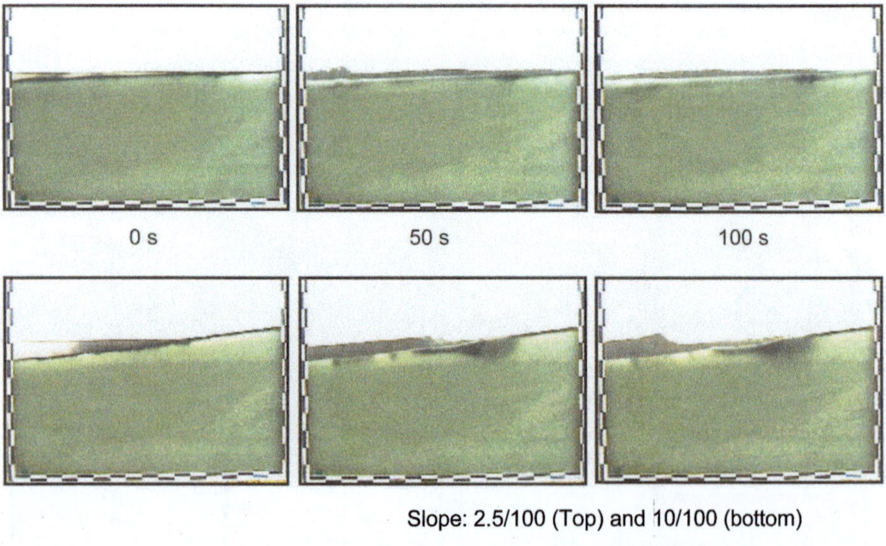

Fig. 2.6. Comparison of flow patterns of particulate matters through porous medium

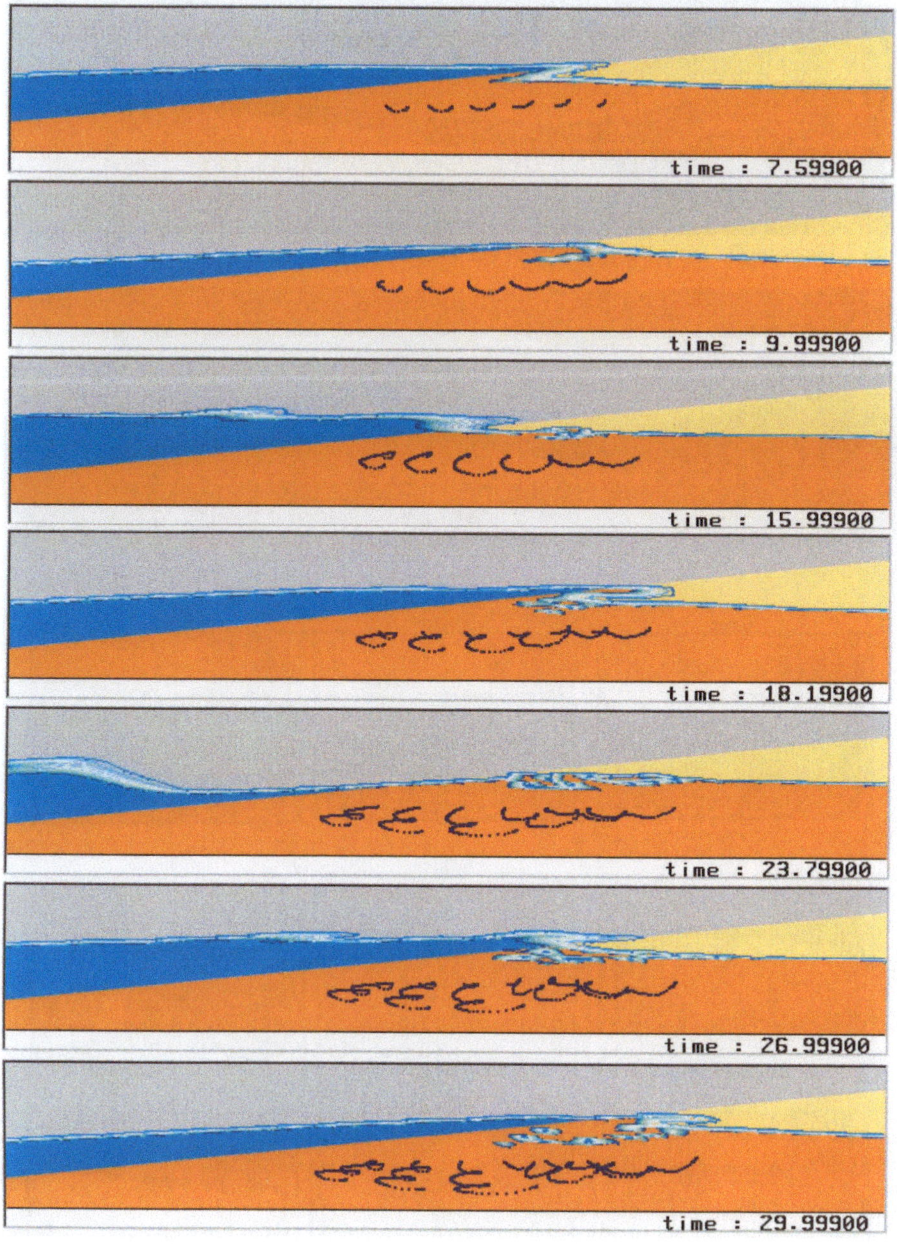

Fig. 3.6. Streak lines (2D case, numerical simulation)

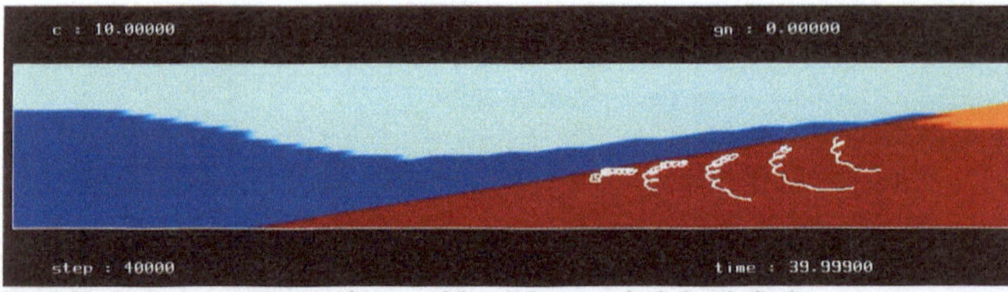

Fig. 3.10. Seawater penetration into sand (g_n=0.0, numerical simulation)

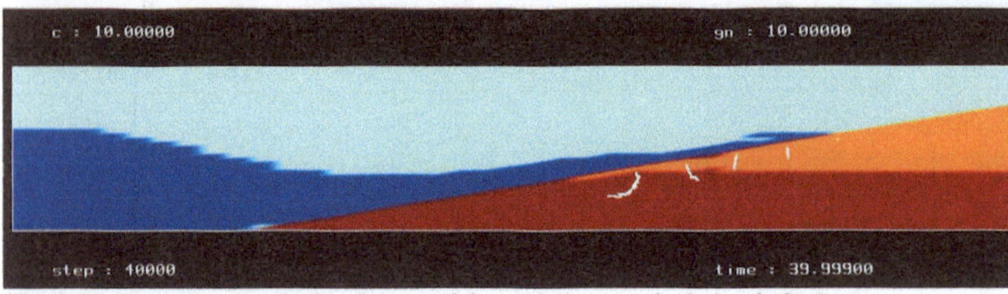

Fig. 3.11. Seawater penetration into sand (g_n =10.0, numerical simulation)

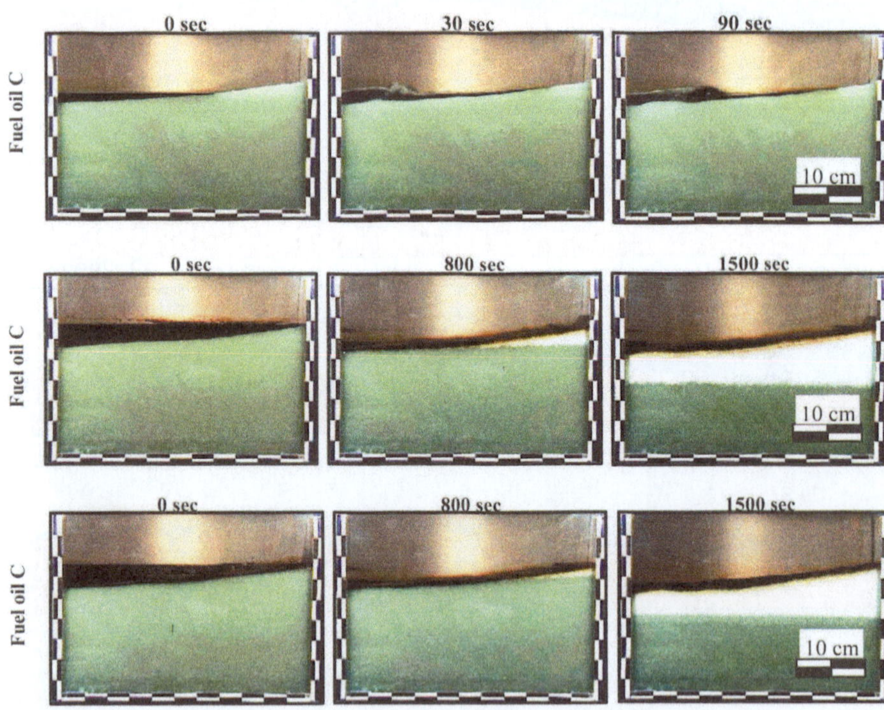

Fig.4.1. Infiltration of fuel oil C into simulated sandy beach (glass beads) by waves and/or tides

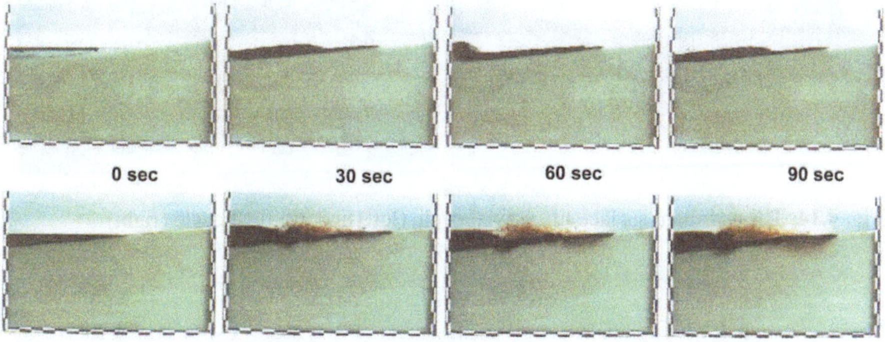

Fig. 4.8. Behavior of weathered oil (top) and dispersed oil (bottom) on sandy beach by wave actions

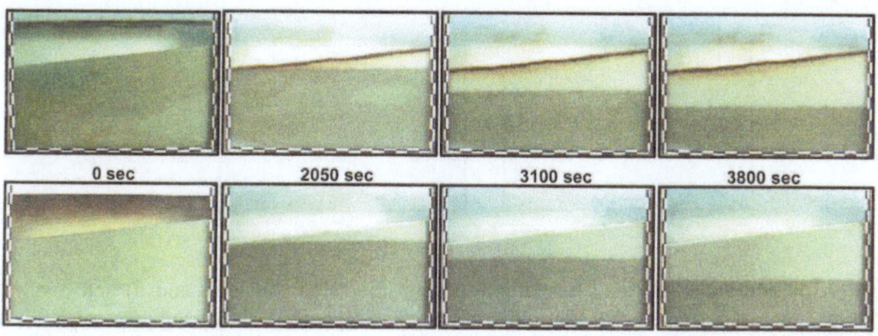

Fig. 4.9. Behavior of weathered oil (top) and dispersed oil (bottom) on sandy beach by tide

Fig. 4.10. Behavior of weathered oil (top) and dispersed oil (bottom) by wave and tidal actions

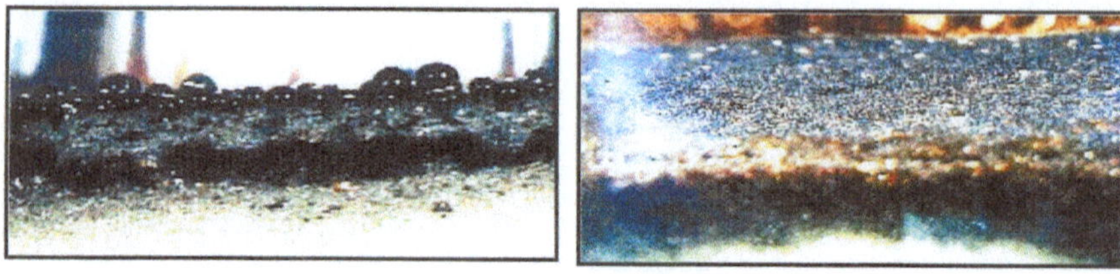

Fig. 4.14. The sediment surface in 1st tidal cycle (left) and 4th tidal cycle (right)

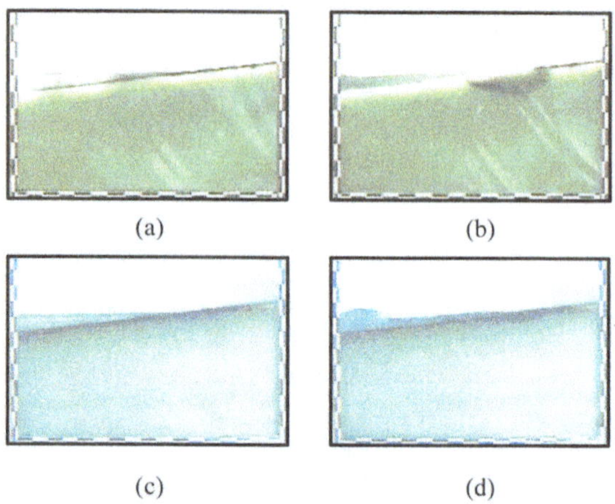

Fig. 4.18. Settling of particulate matters through the oiled (photo, c and d) and unoiled (photo; a and b) porous media. Elapsed times are 0 sec (photo; a and c) and 60 sec (photo; b and d)

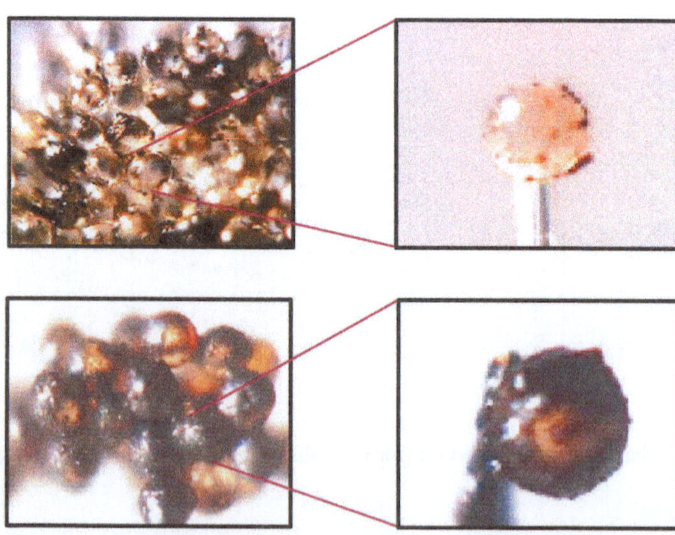

Fig. 4.19. Adhesion of particulate matters on the glass beads coated by crude oil (top) and fuel oil C (bottom). The diameter of glass bead is 1mm.

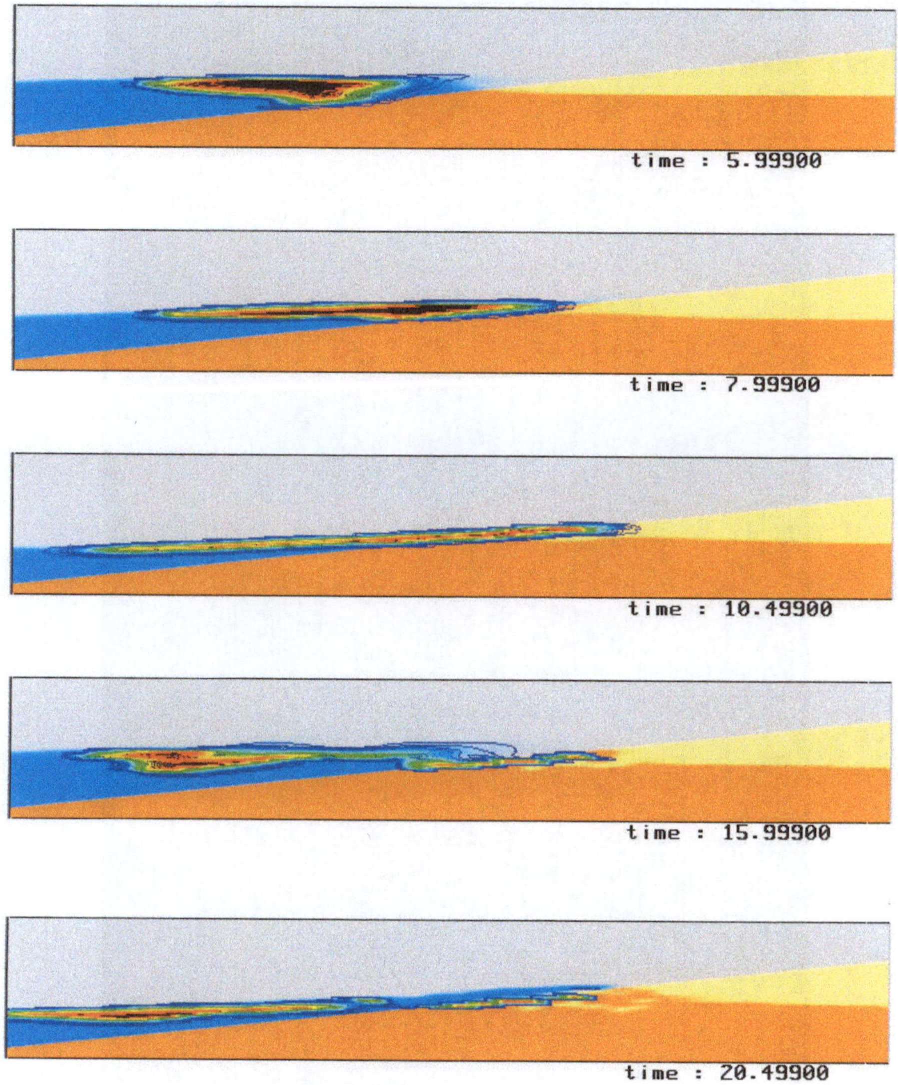

Fig. 5.3. Motion of spilled oil due to wave breaking action (2D case, numerical simulation)

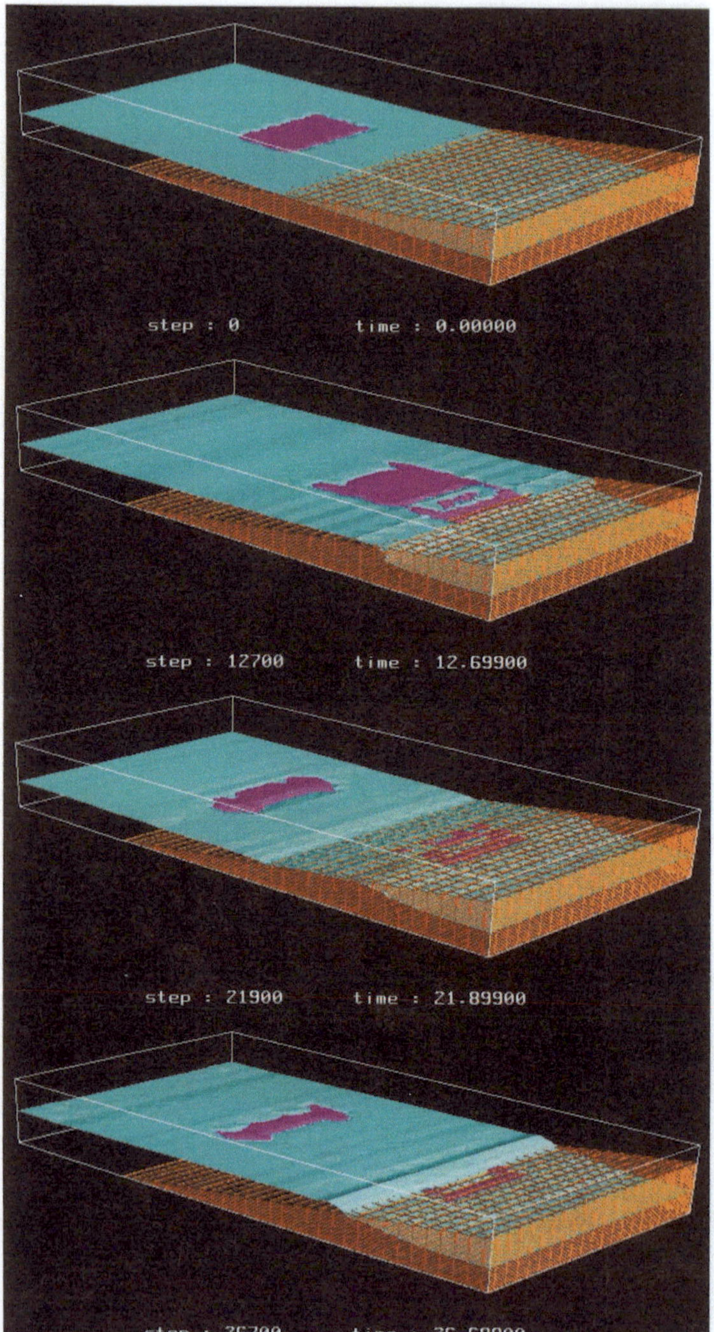

Fig. 5.4. Motion of spilled oil due to wave breaking action (3D case, numerical simulation)

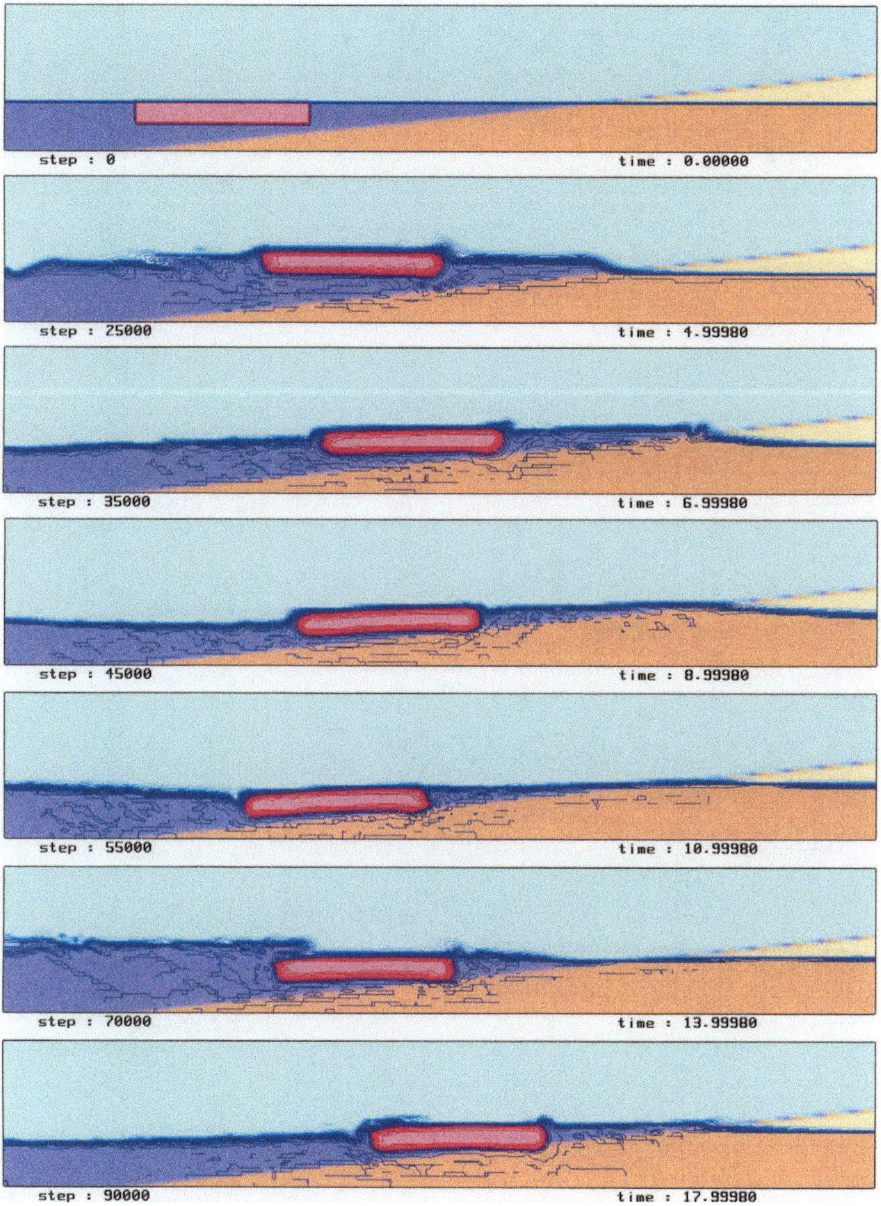

Fig. 5.5. Oil deformation ($\mu^0_{oil} = 10^4$) (2D case, numerical simulation)

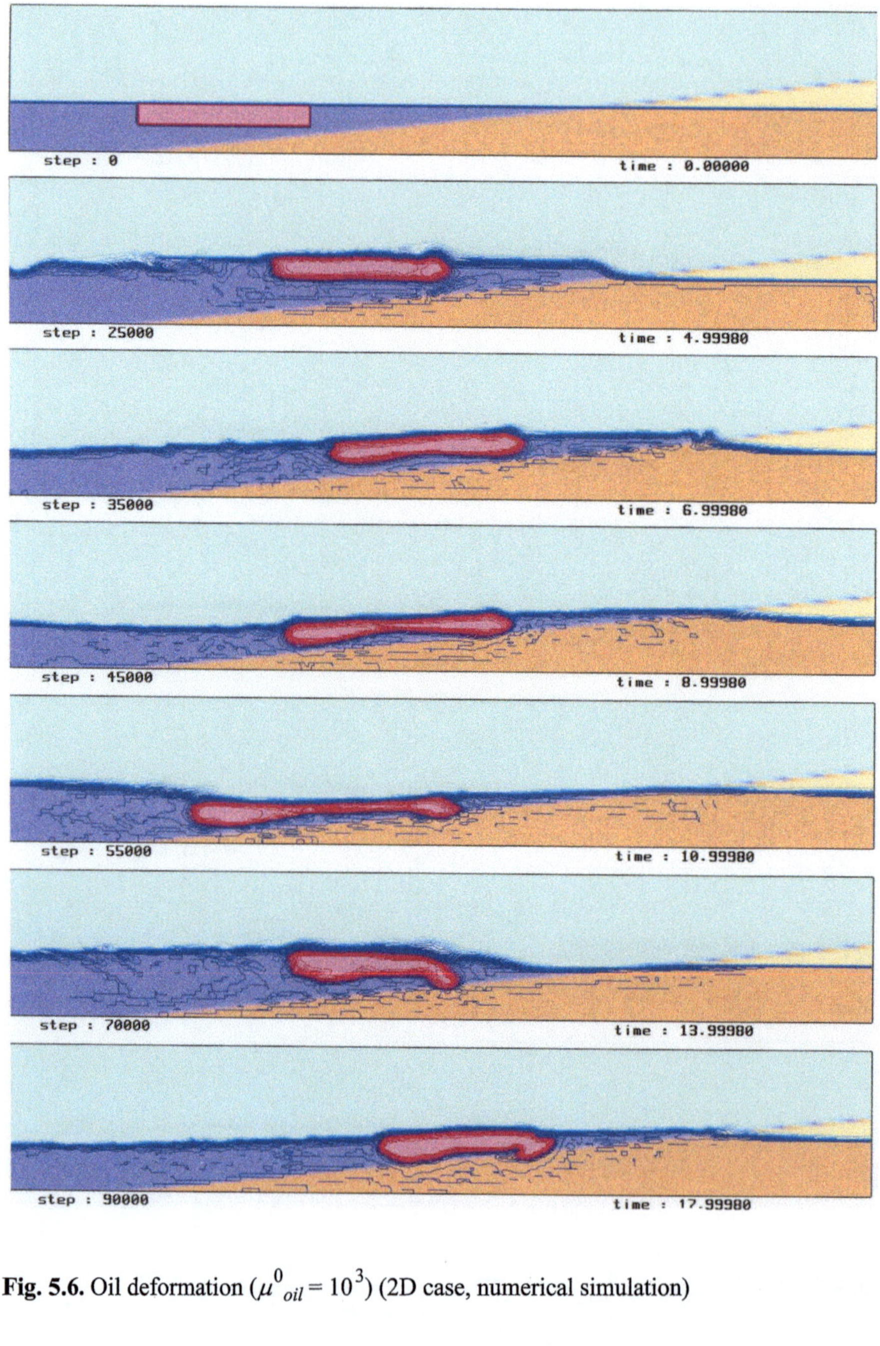

Fig. 5.6. Oil deformation ($\mu^0_{oil} = 10^3$) (2D case, numerical simulation)

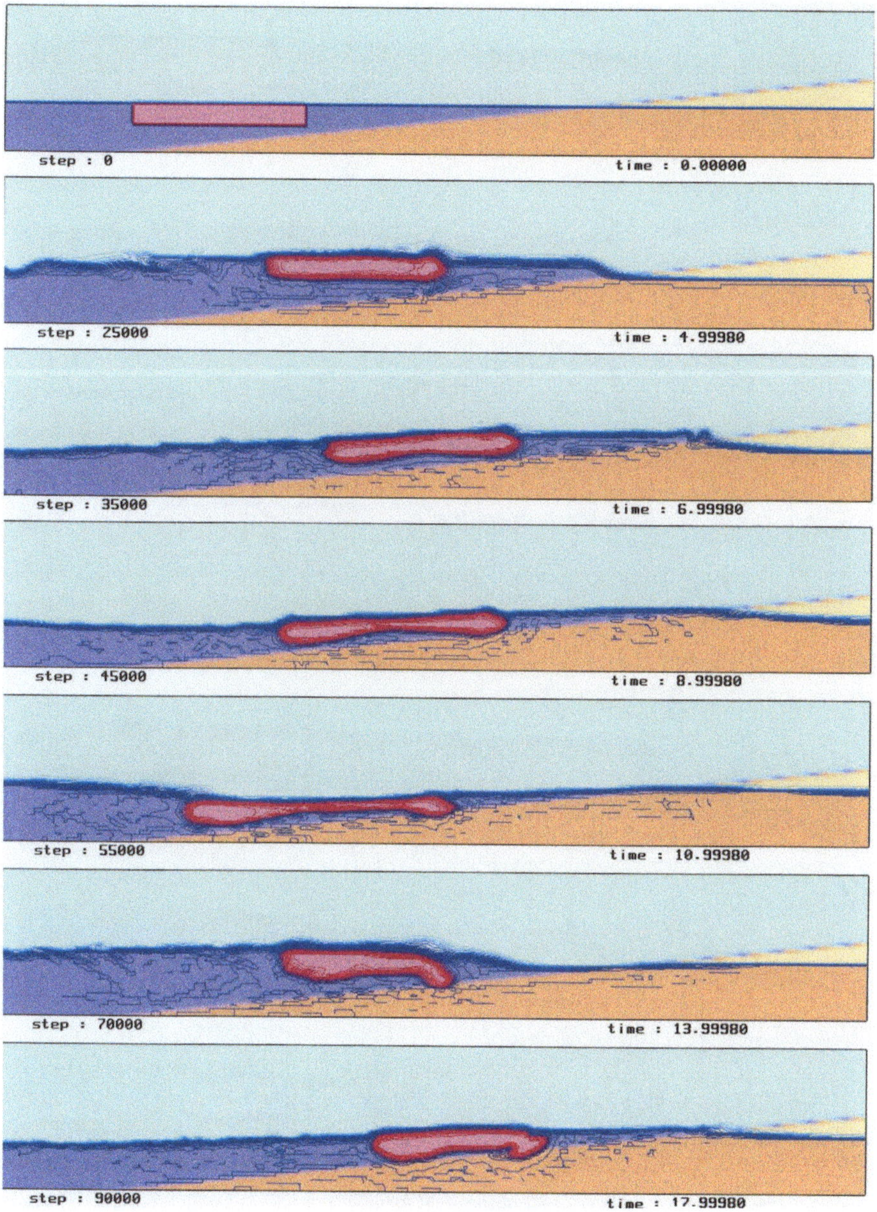

Fig. 5.7. Oil deformation ($\mu^{0}_{oil} = 10^{2}$) (2D case, numerical simulation)

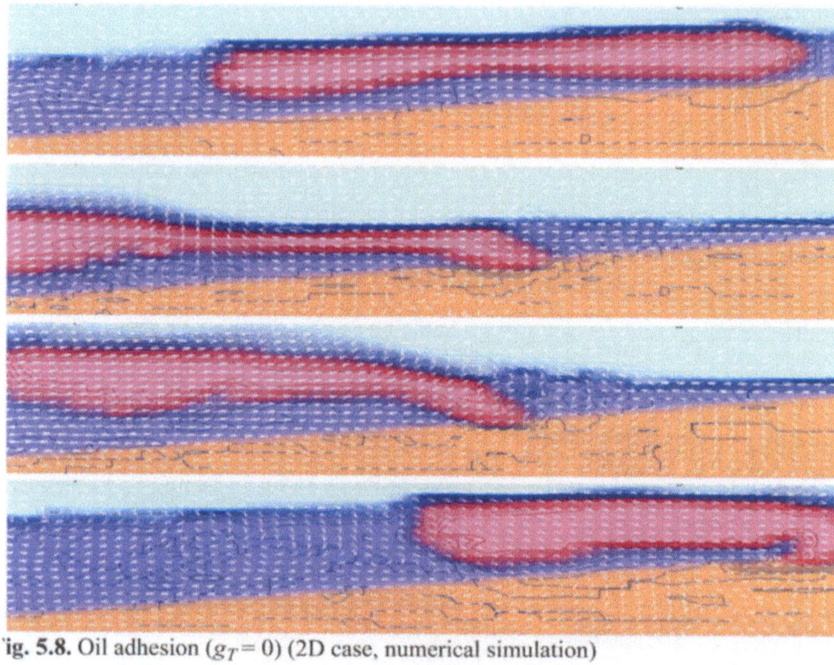

'ig. **5.8.** Oil adhesion ($g_T = 0$) (2D case, numerical simulation)

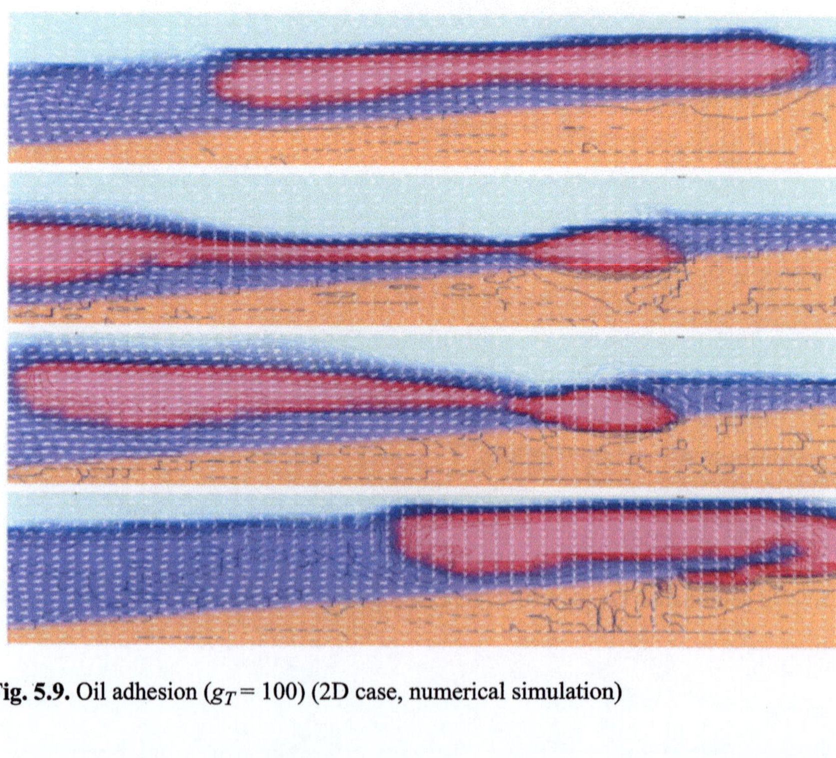

Fig. 5.9. Oil adhesion ($g_T = 100$) (2D case, numerical simulation)

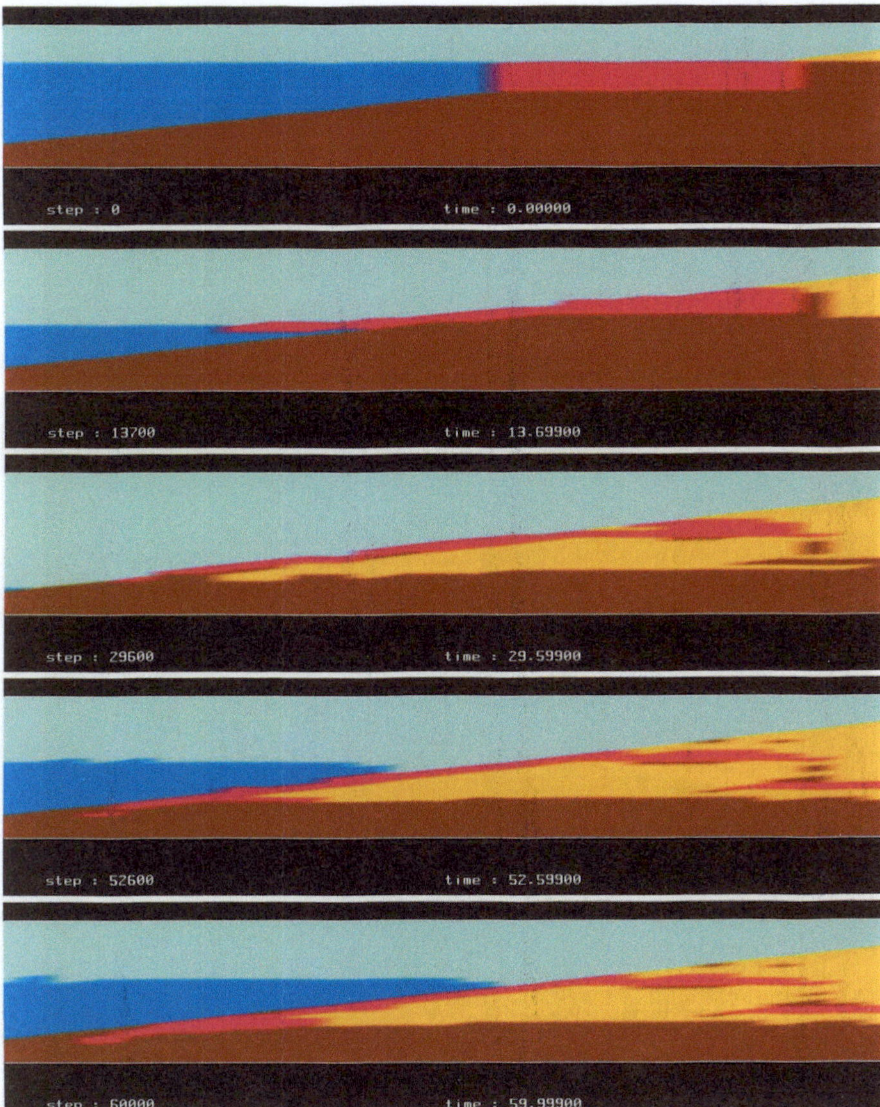

Fig. 5.11. Motion of spilled oil due to tidal motion (2D case, numerical simulation)

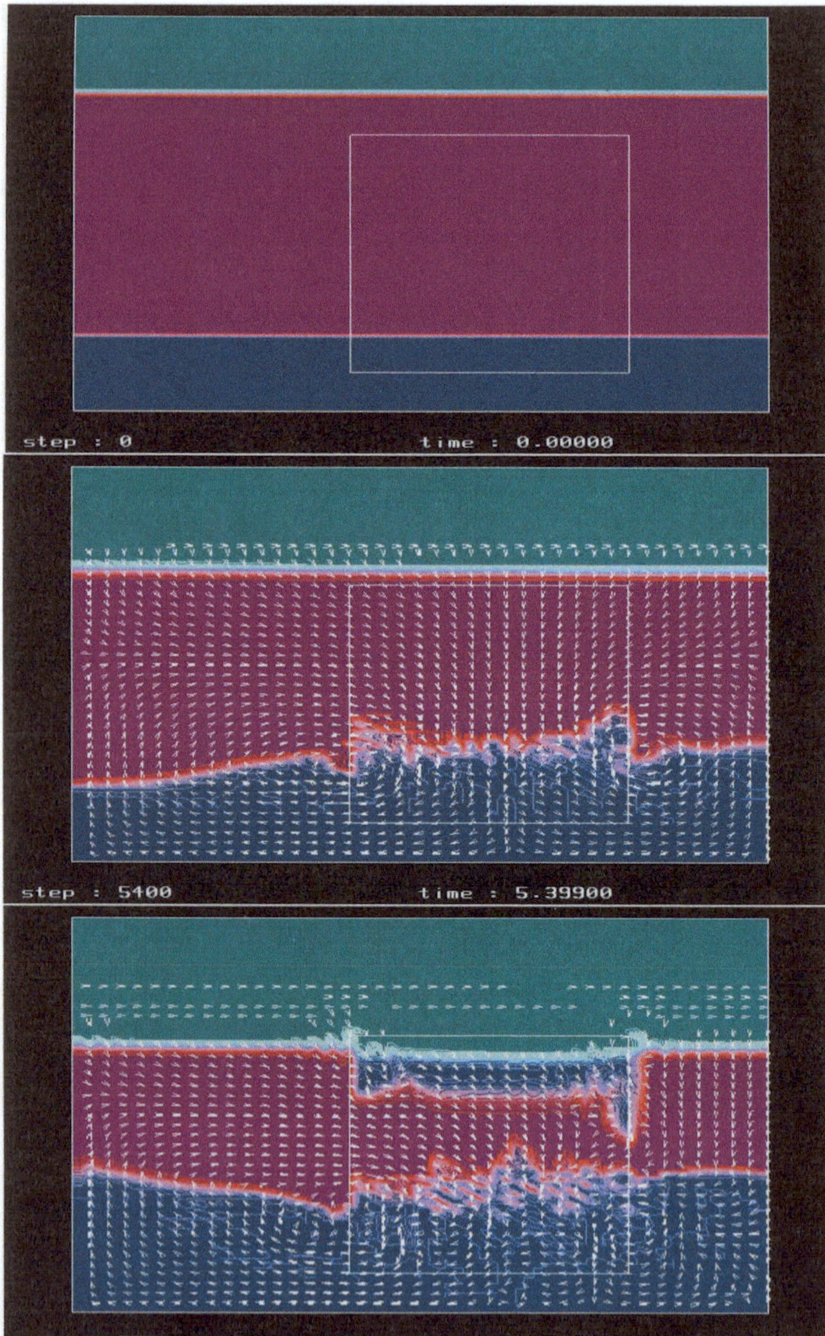

Fig. 7.2. Decomposition of oil (2D case, numerical simulation)

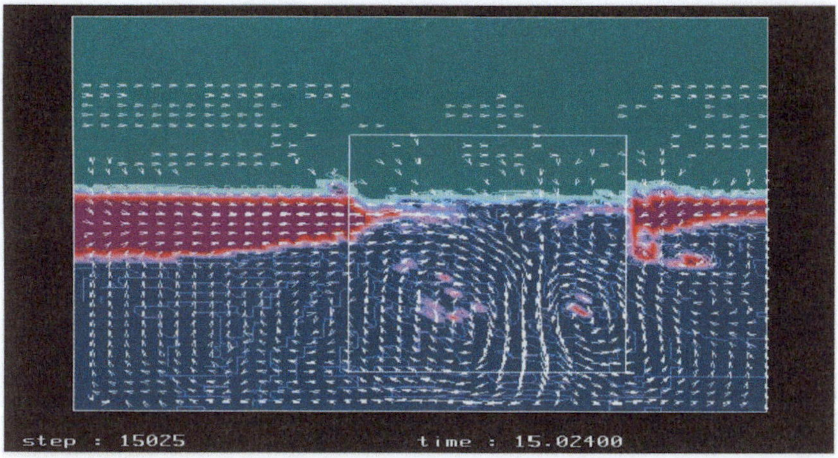

Fig. 7.2. Decomposition of oil (2D case, numerical simulation, continued)

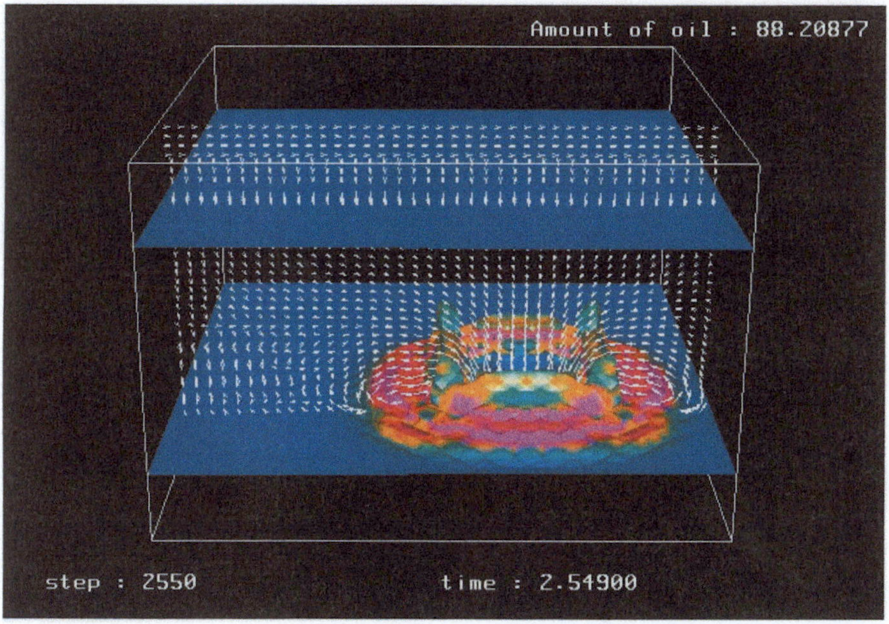

Fig. 7.3. Decomposition of oil (3D case, Numerical simulation)

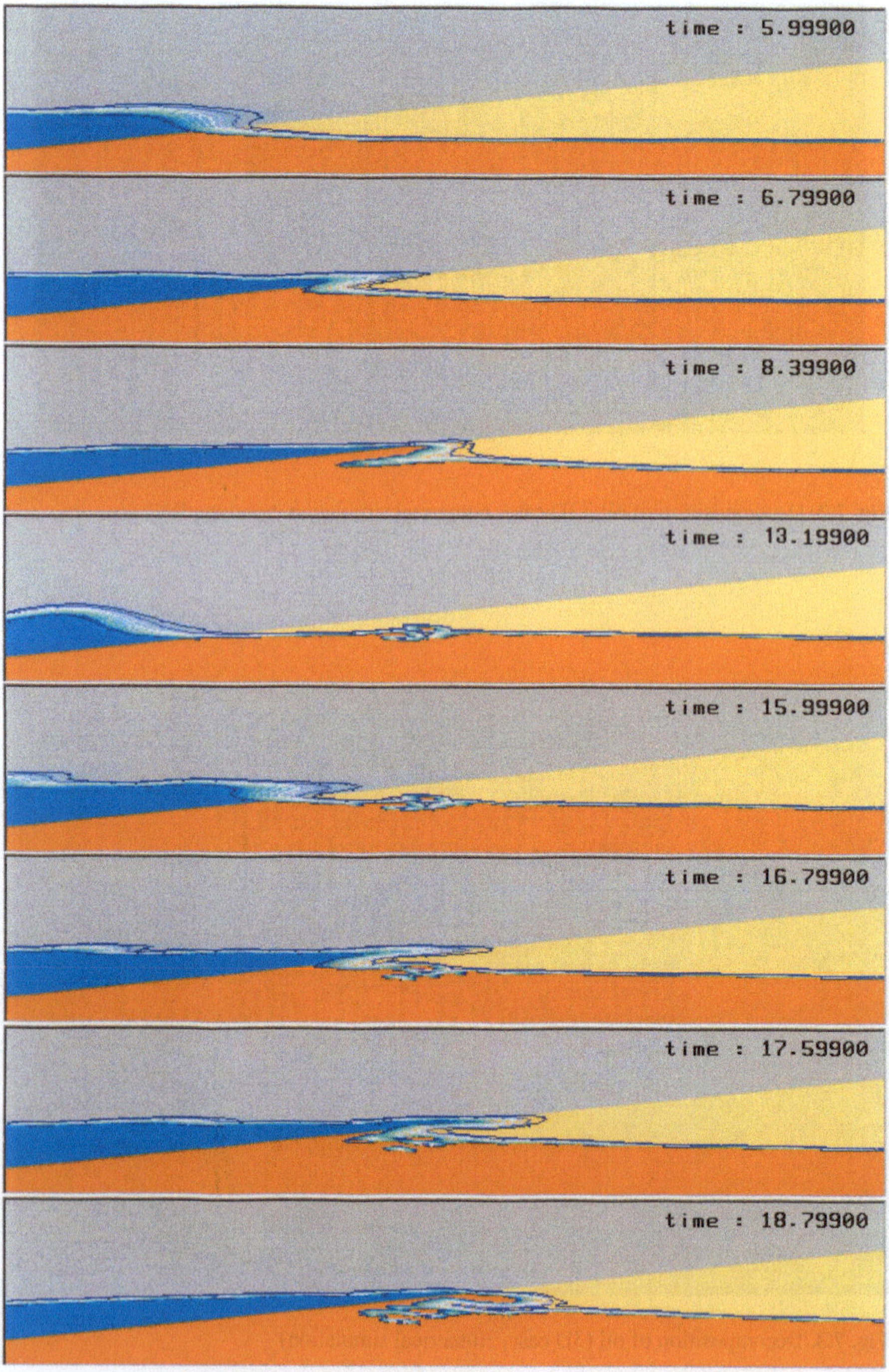

Fig. 8.4. Aeration (2D case, numerical simulation)

Index

MIX
Papier aus verantwortungsvollen Quellen
Paper from responsible sources
FSC® C105338

If you have any concerns about our products,
you can contact us on
ProductSafety@springernature.com

In case Publisher is established outside the EU,
the EU authorized representative is:
Springer Nature Customer Service Center GmbH
Europaplatz 3, 69115 Heidelberg, Germany

Printed by Libri Plureos GmbH
in Hamburg, Germany